Alex Box
by Rankin

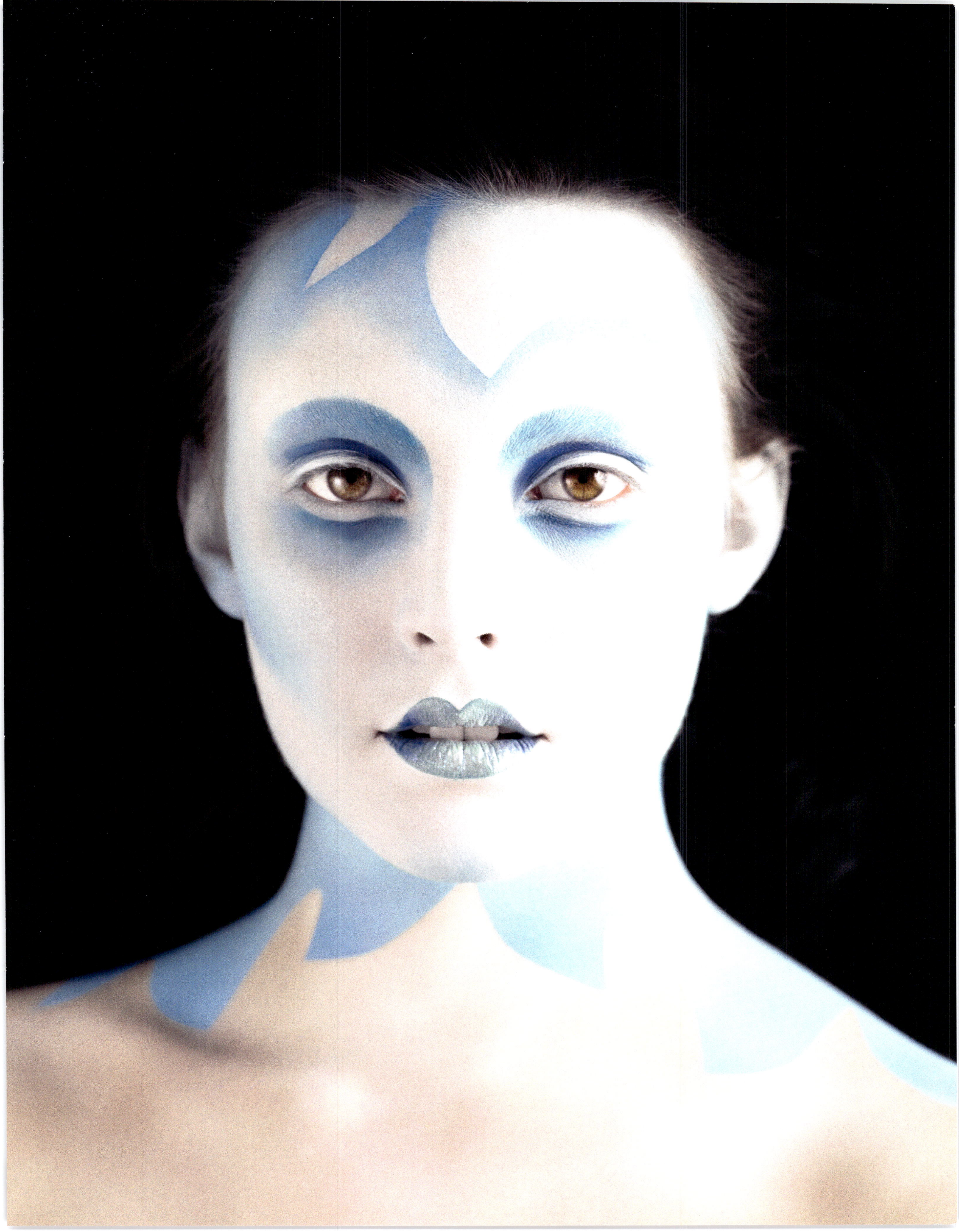

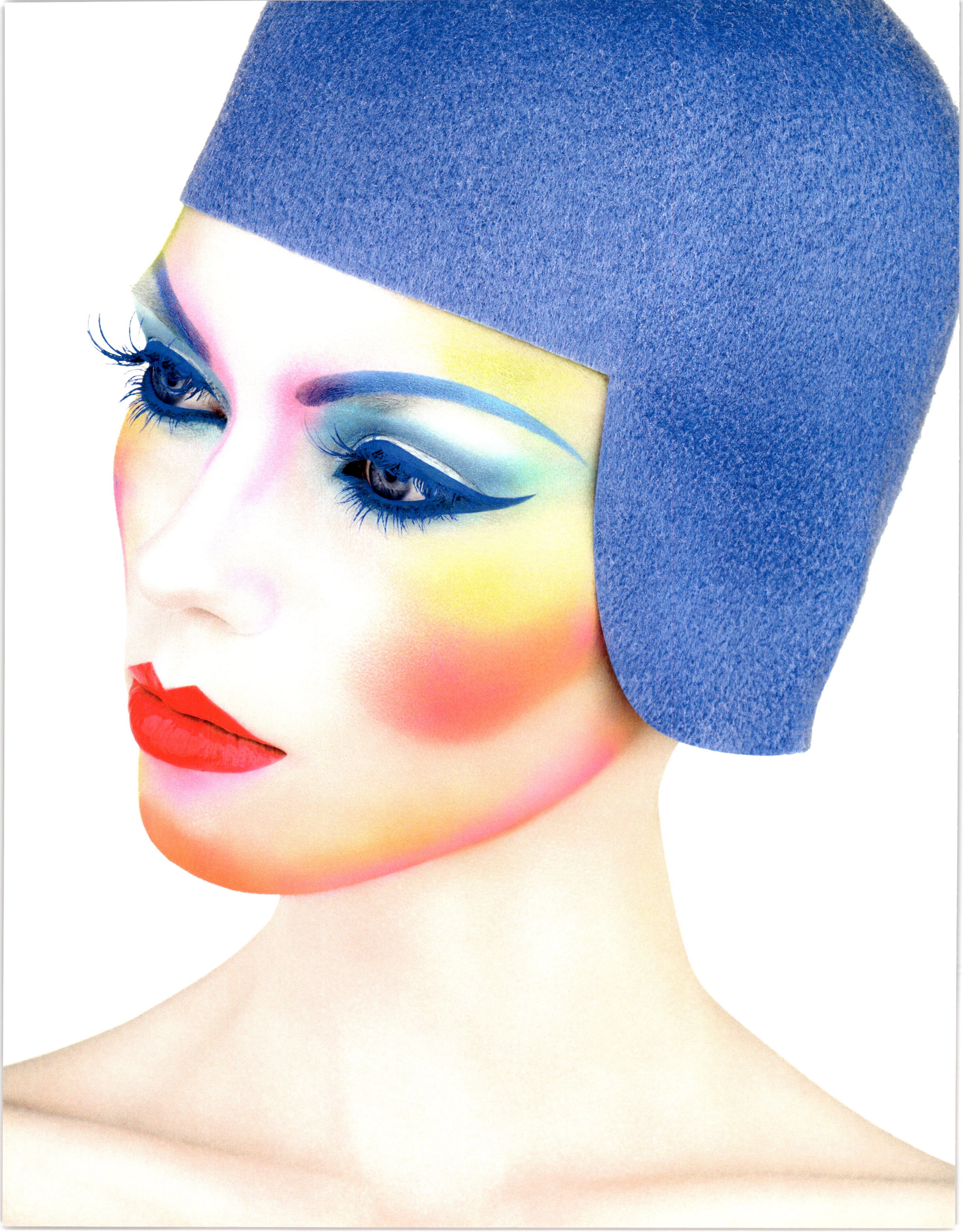

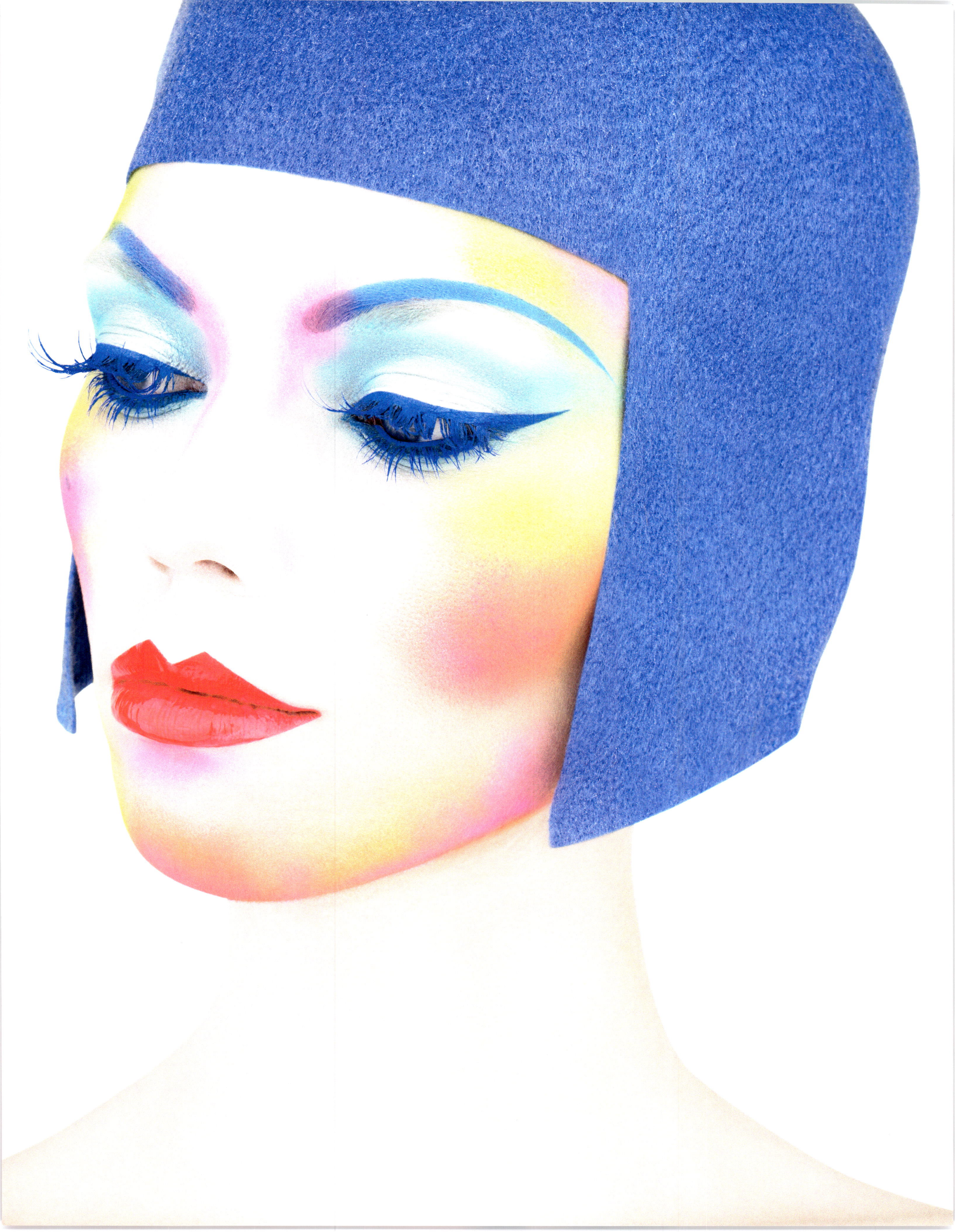

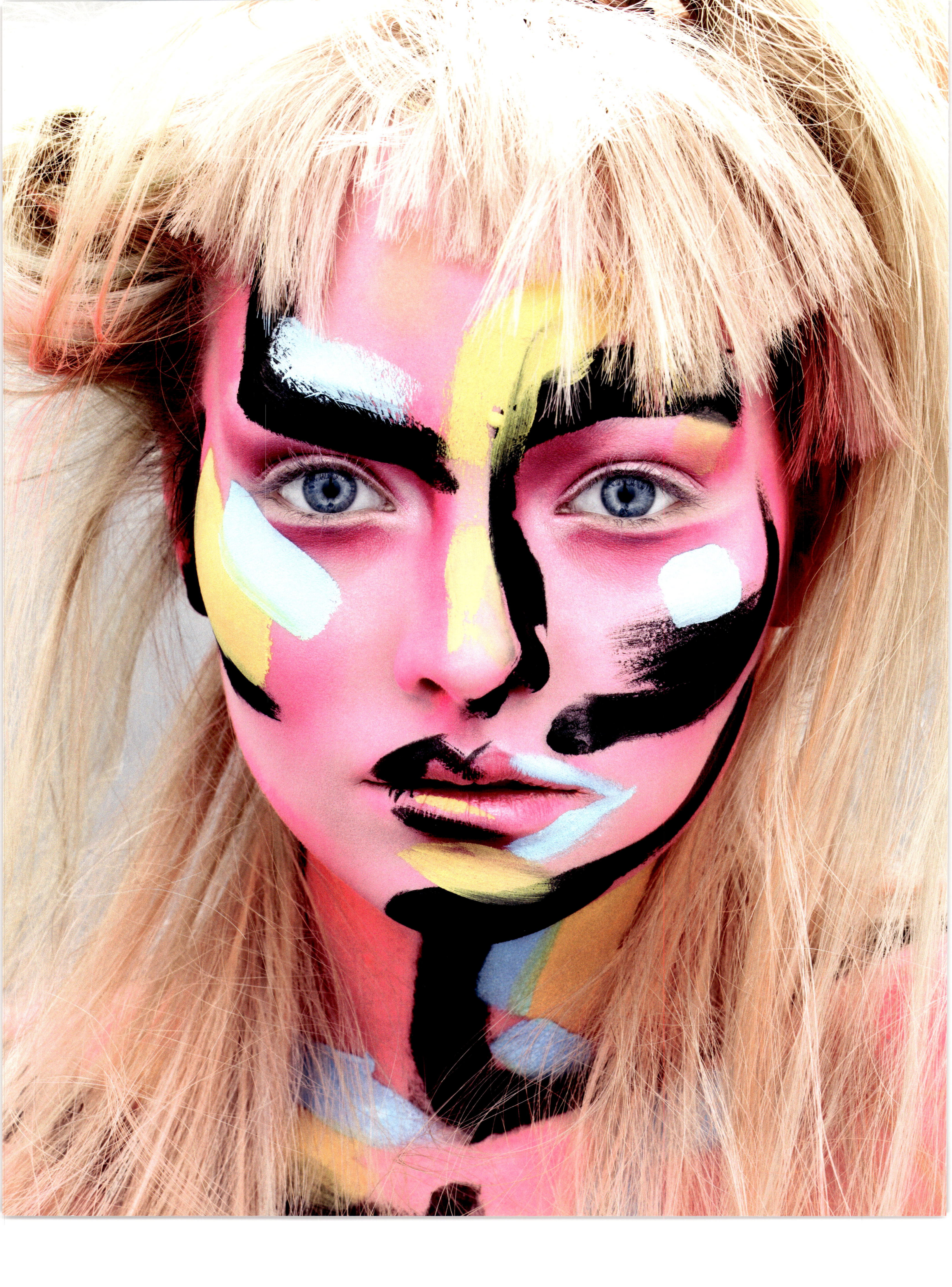

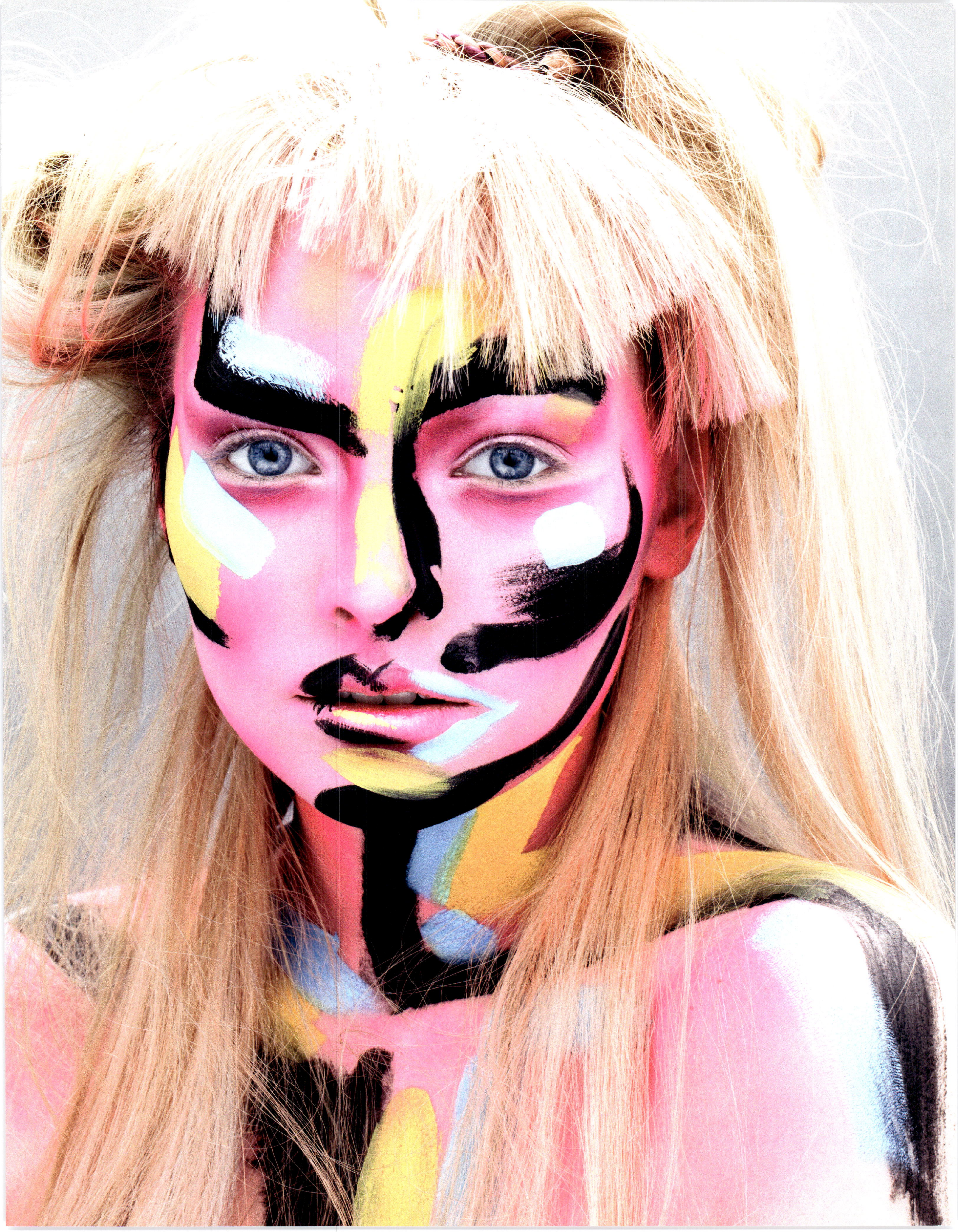

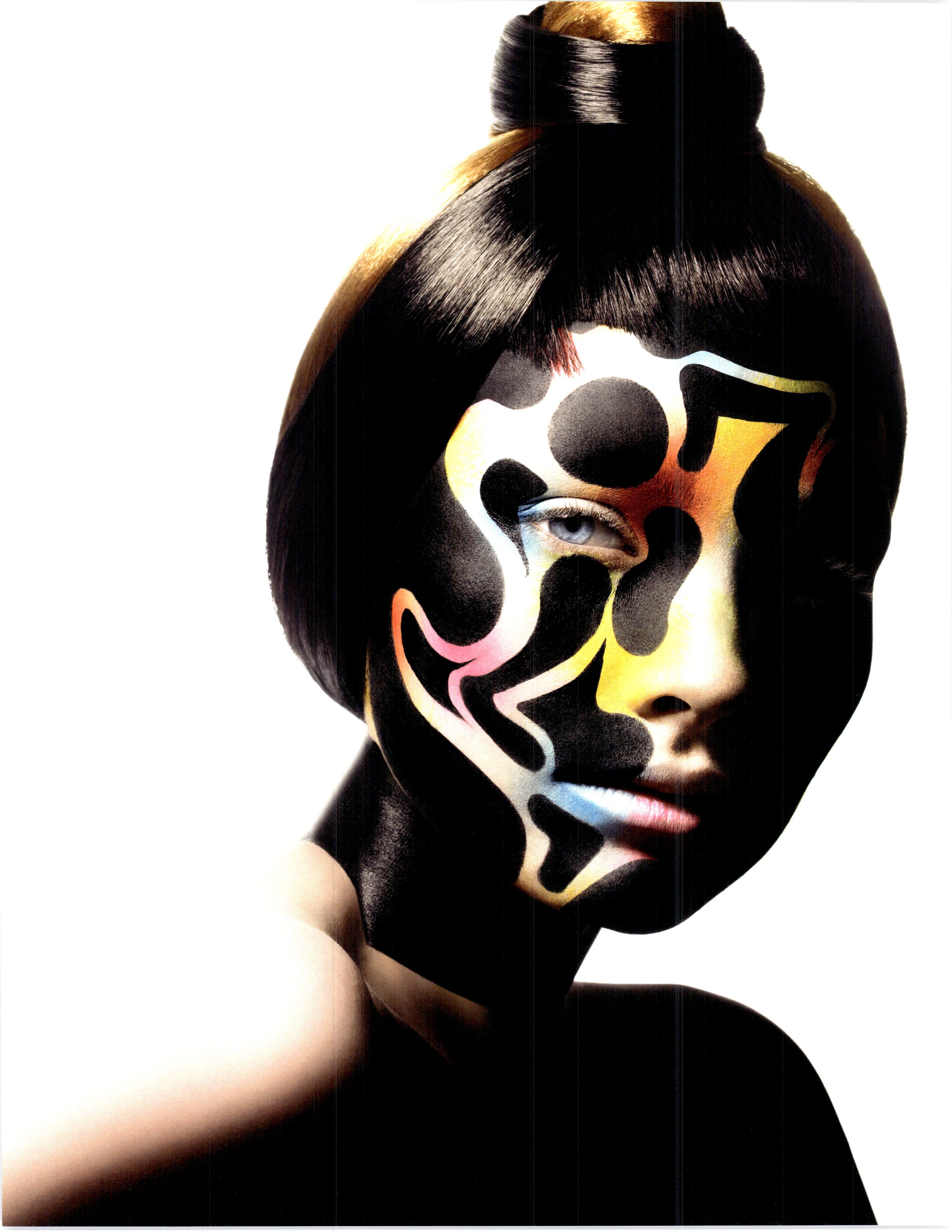

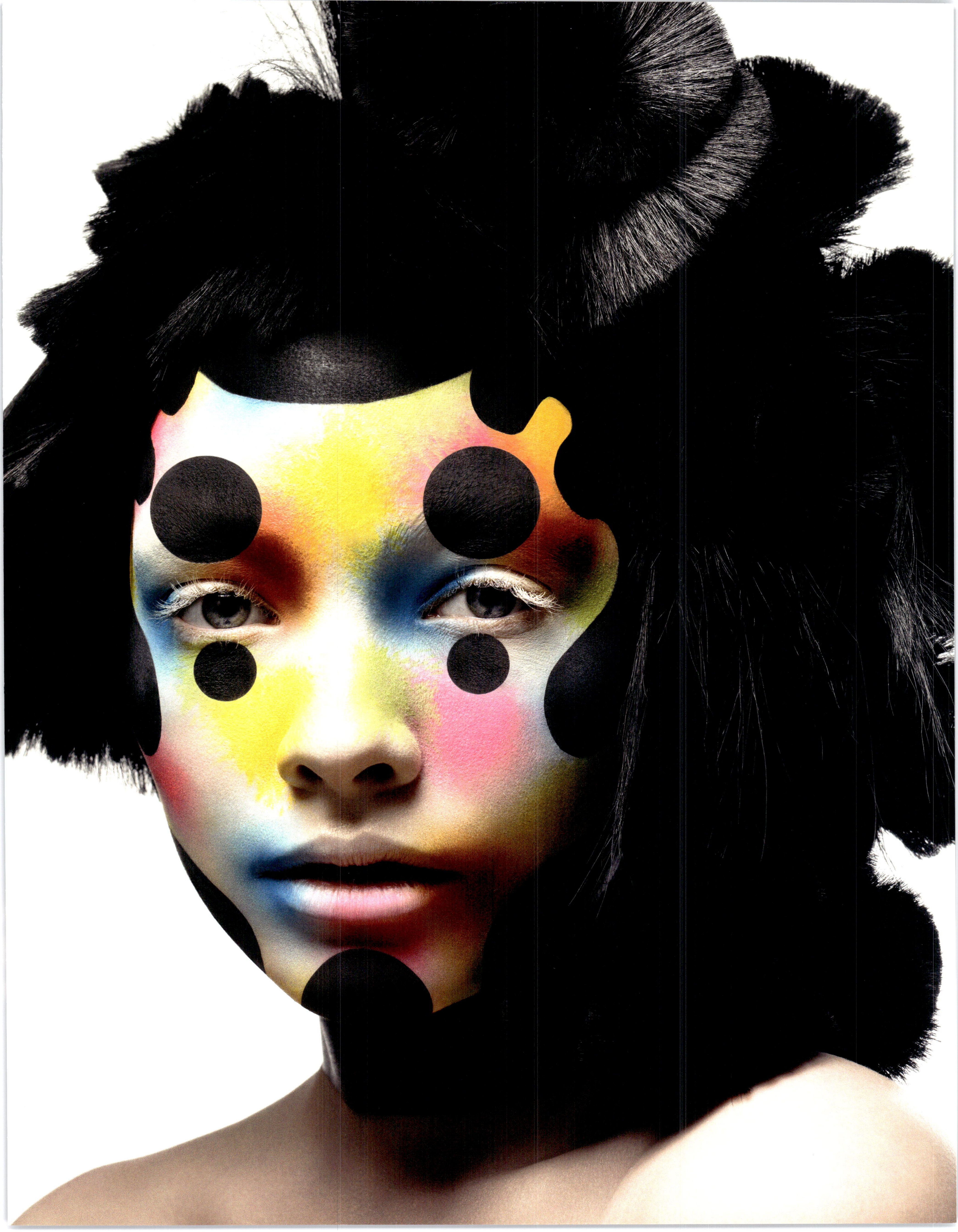

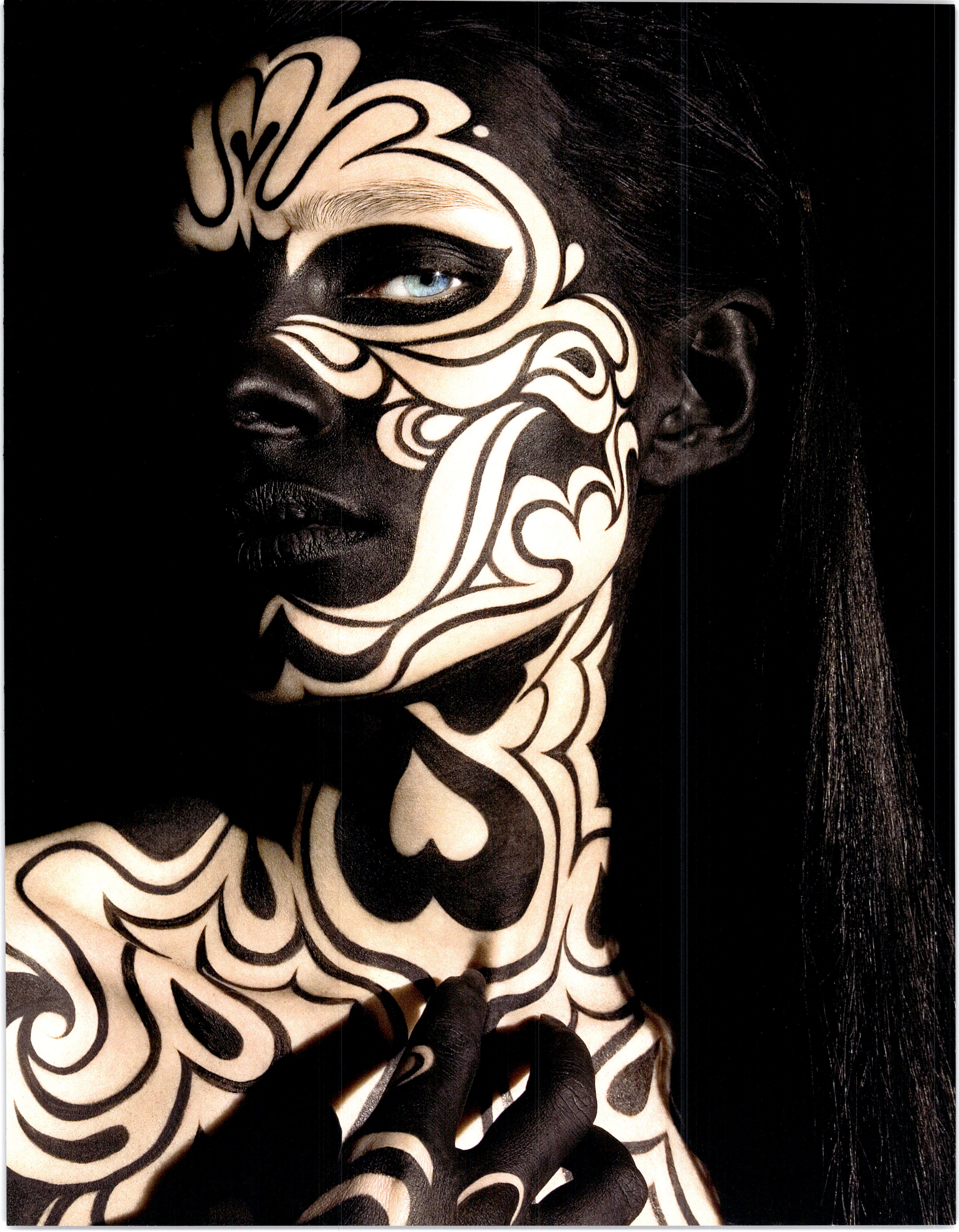

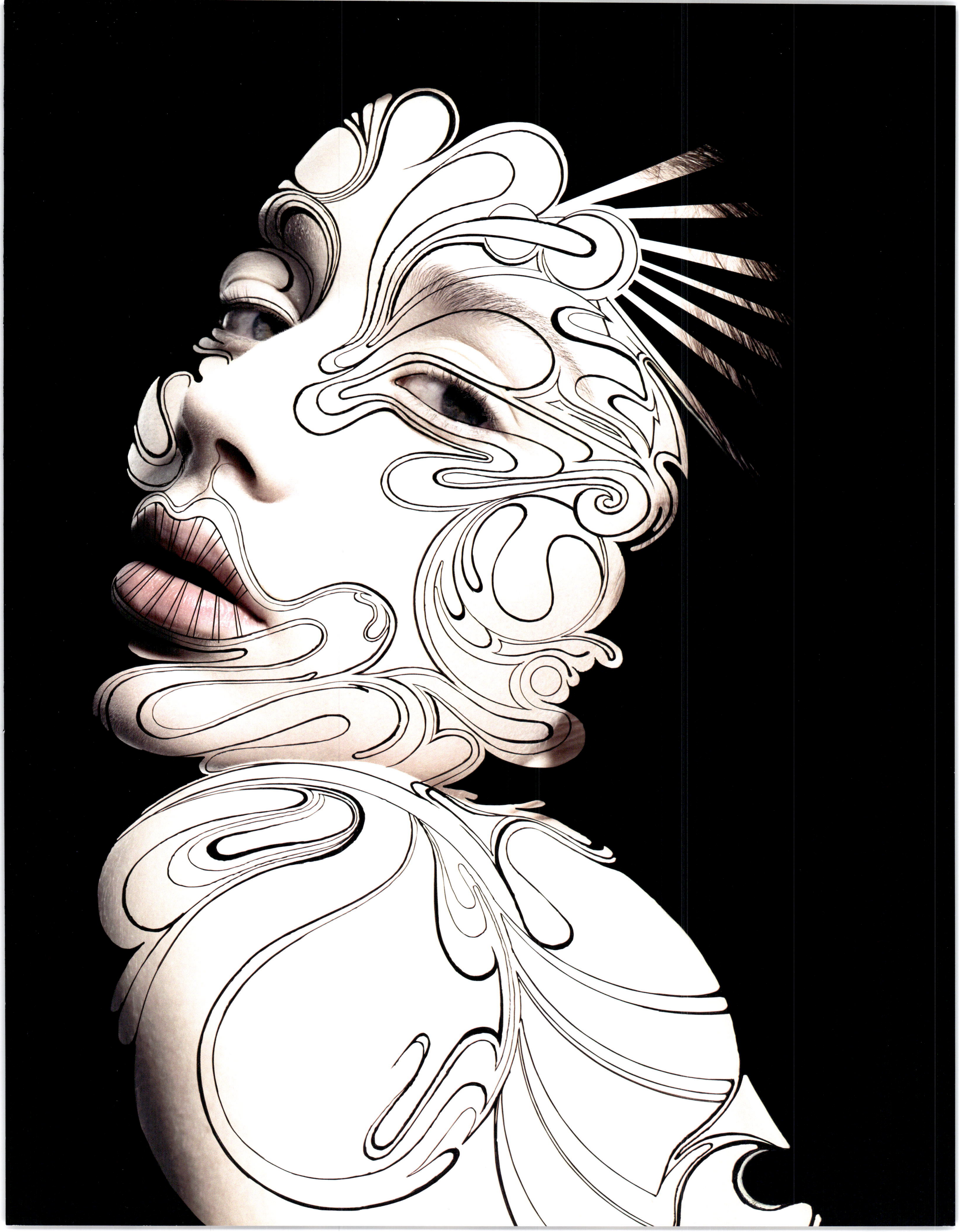

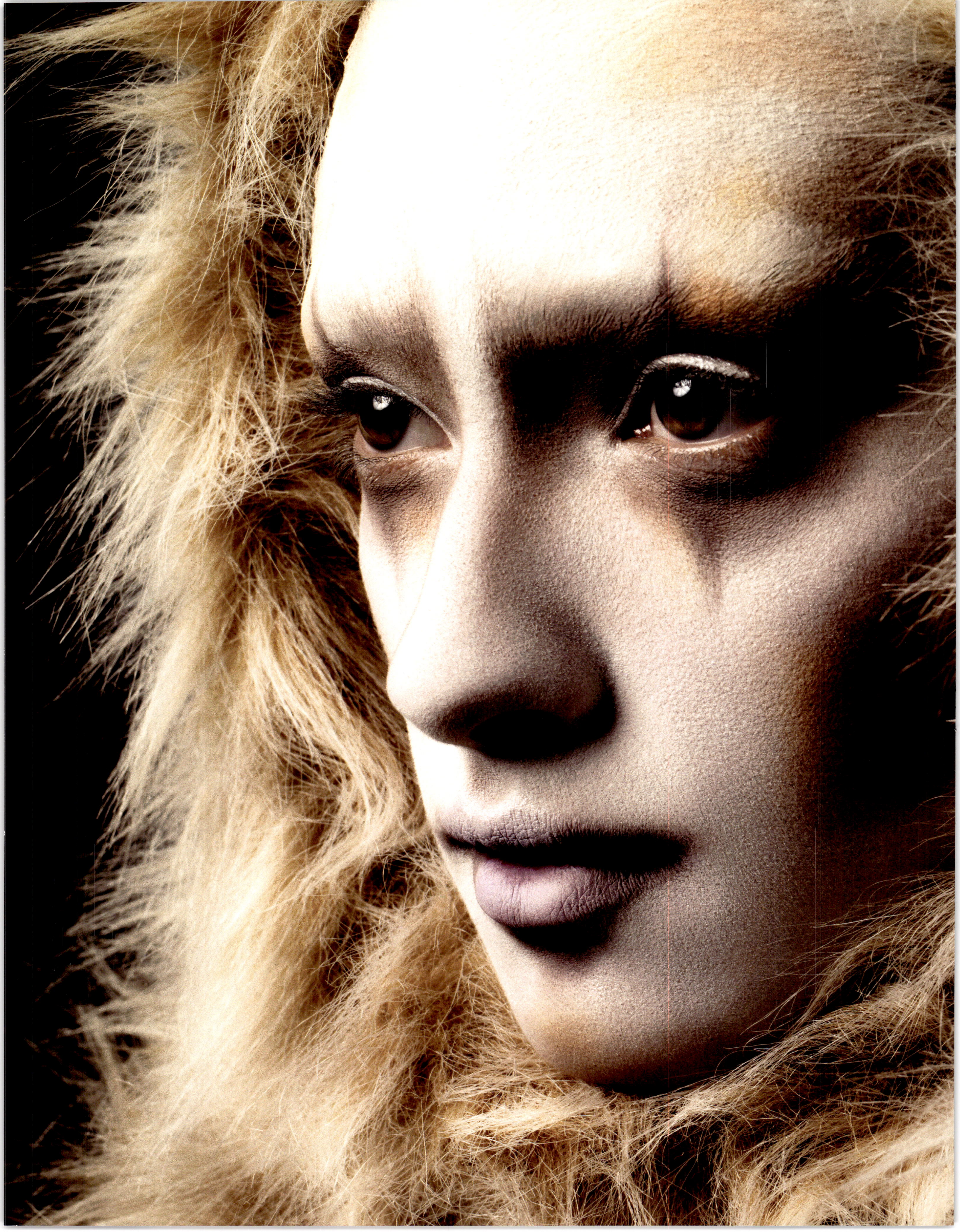

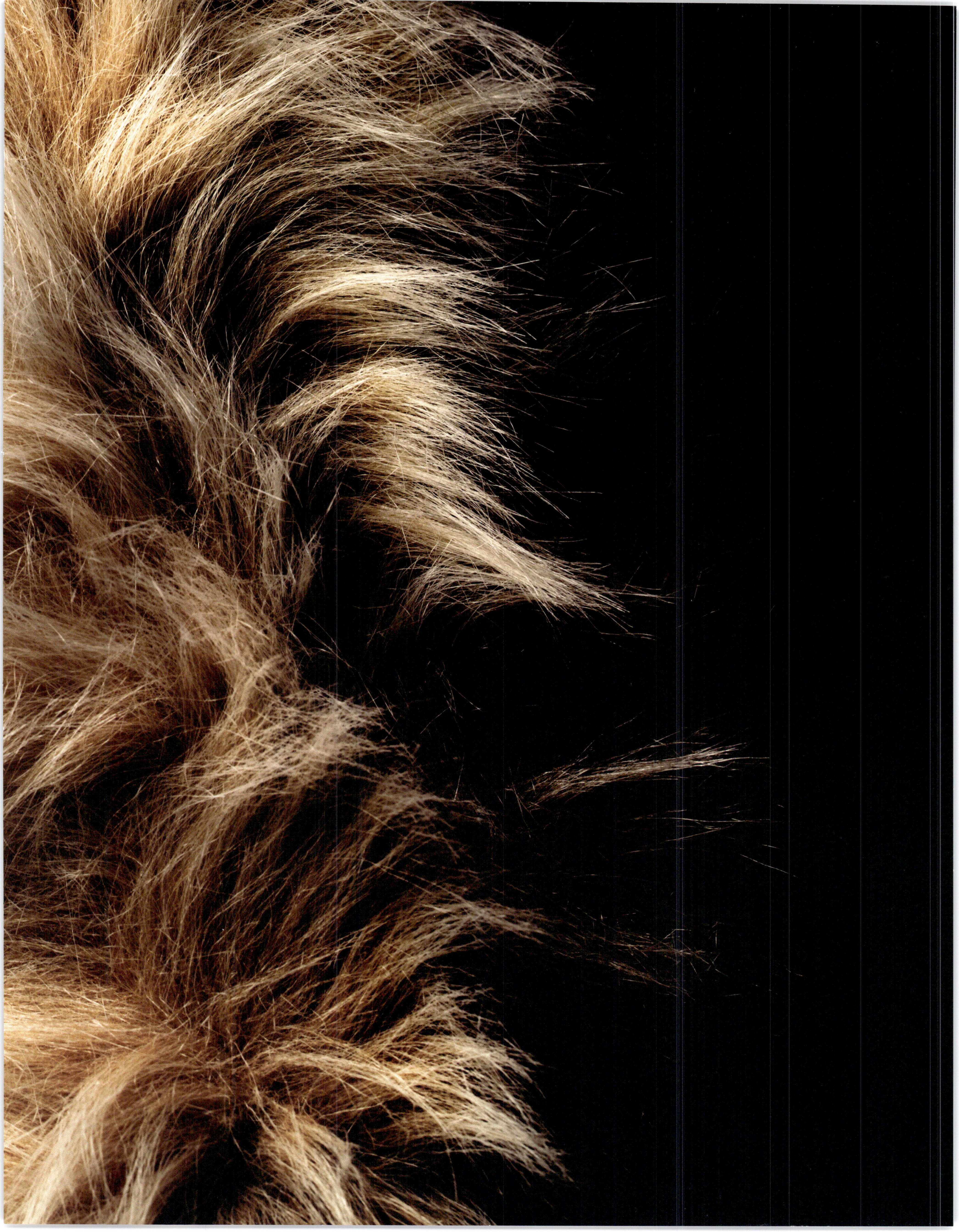

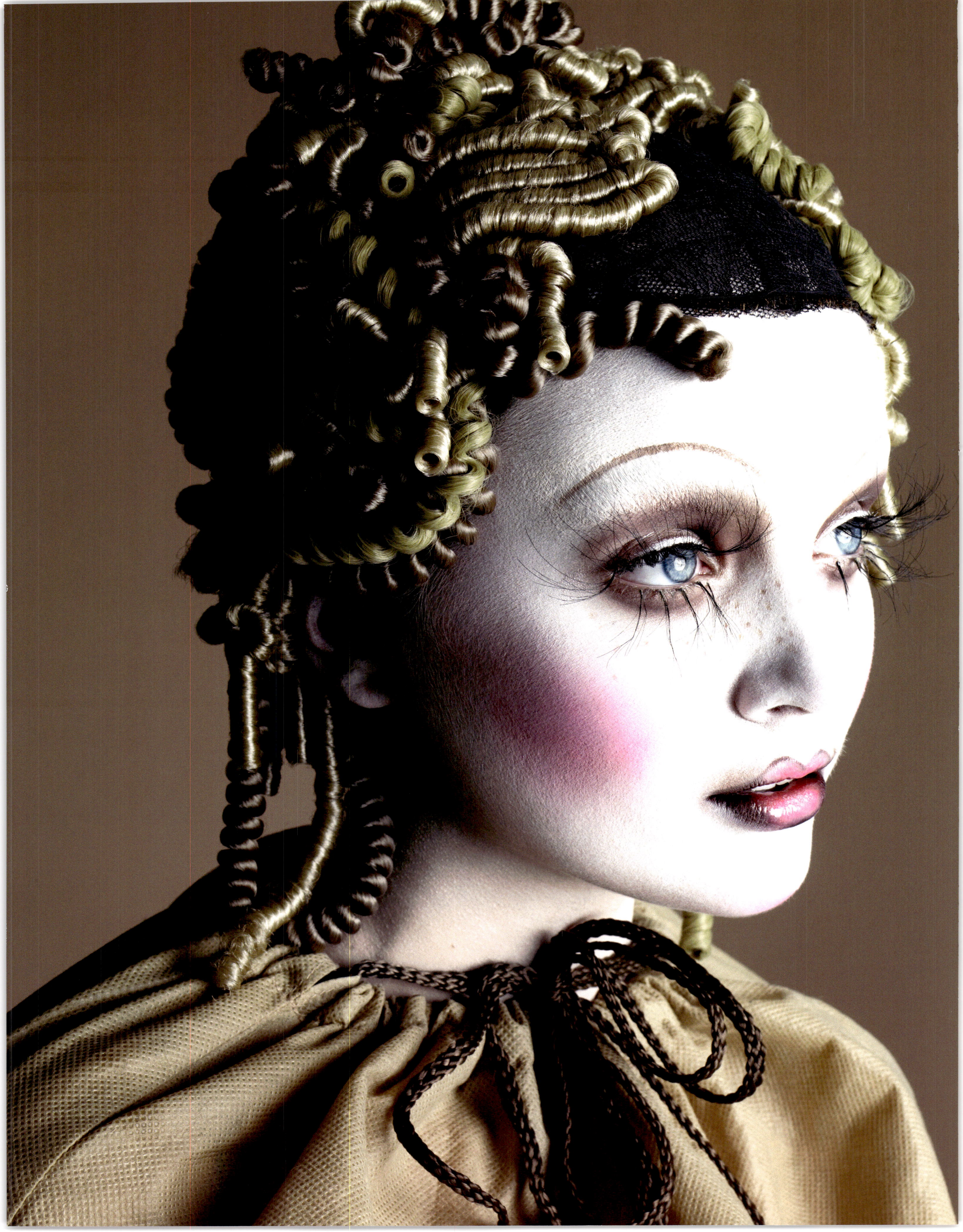

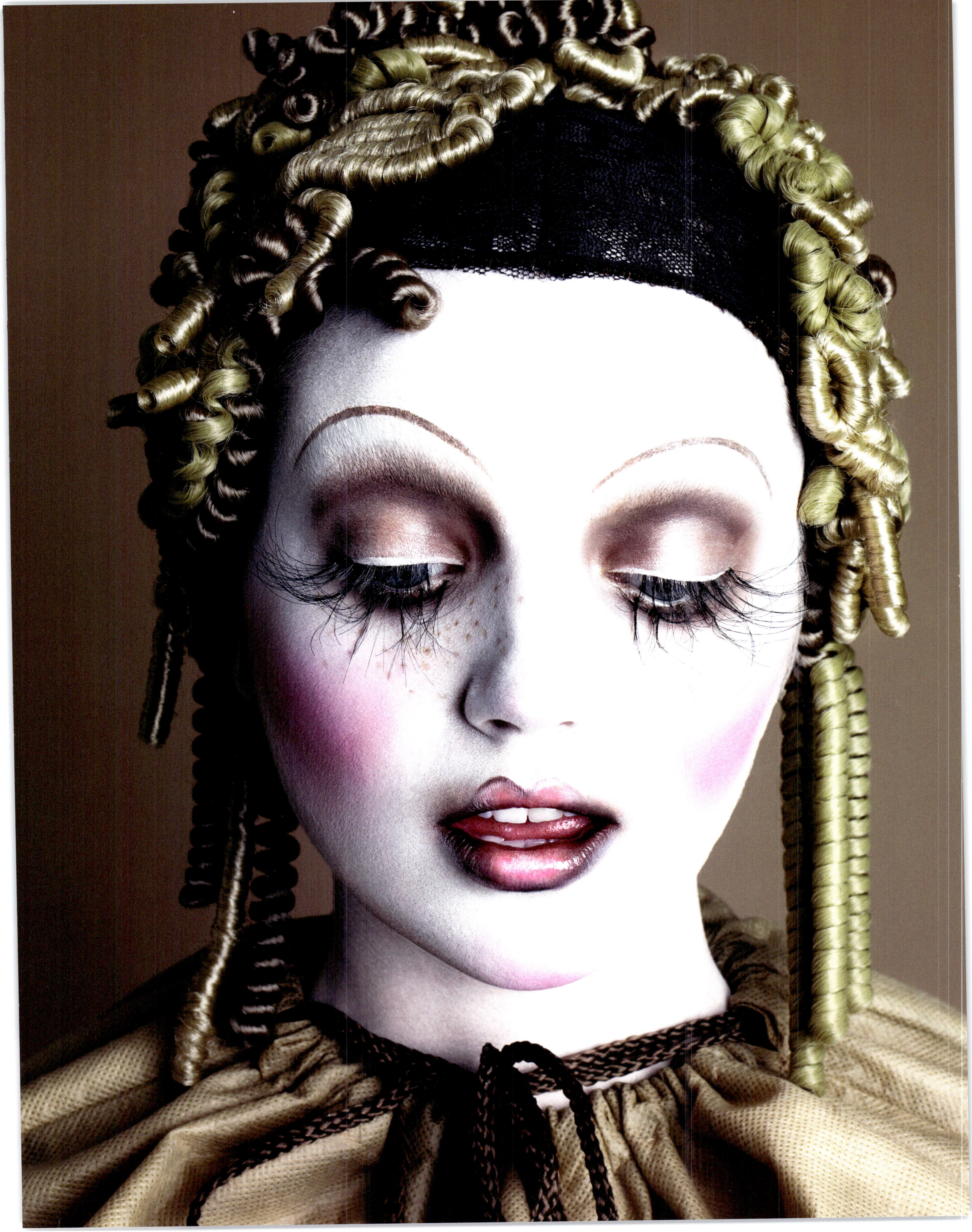

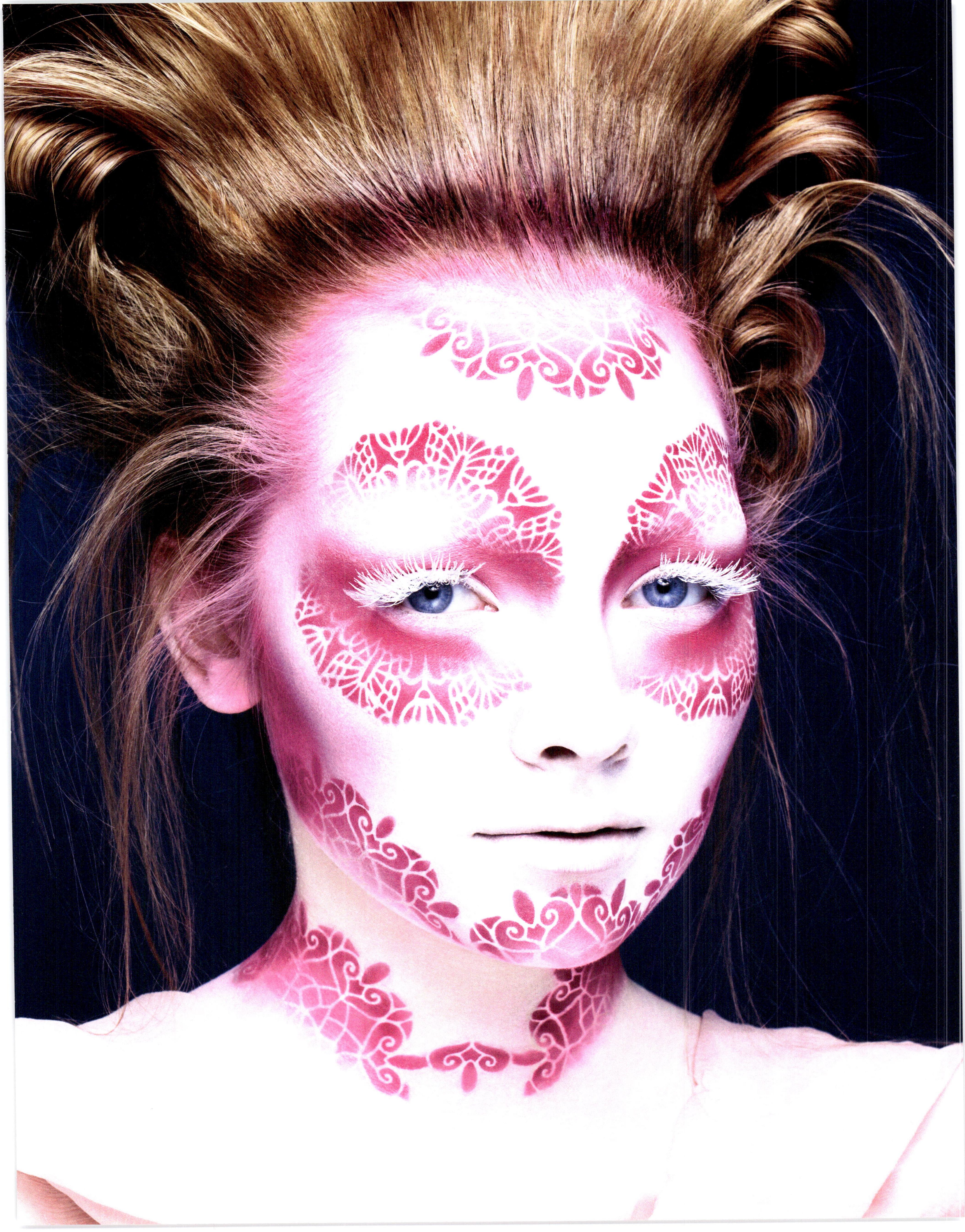

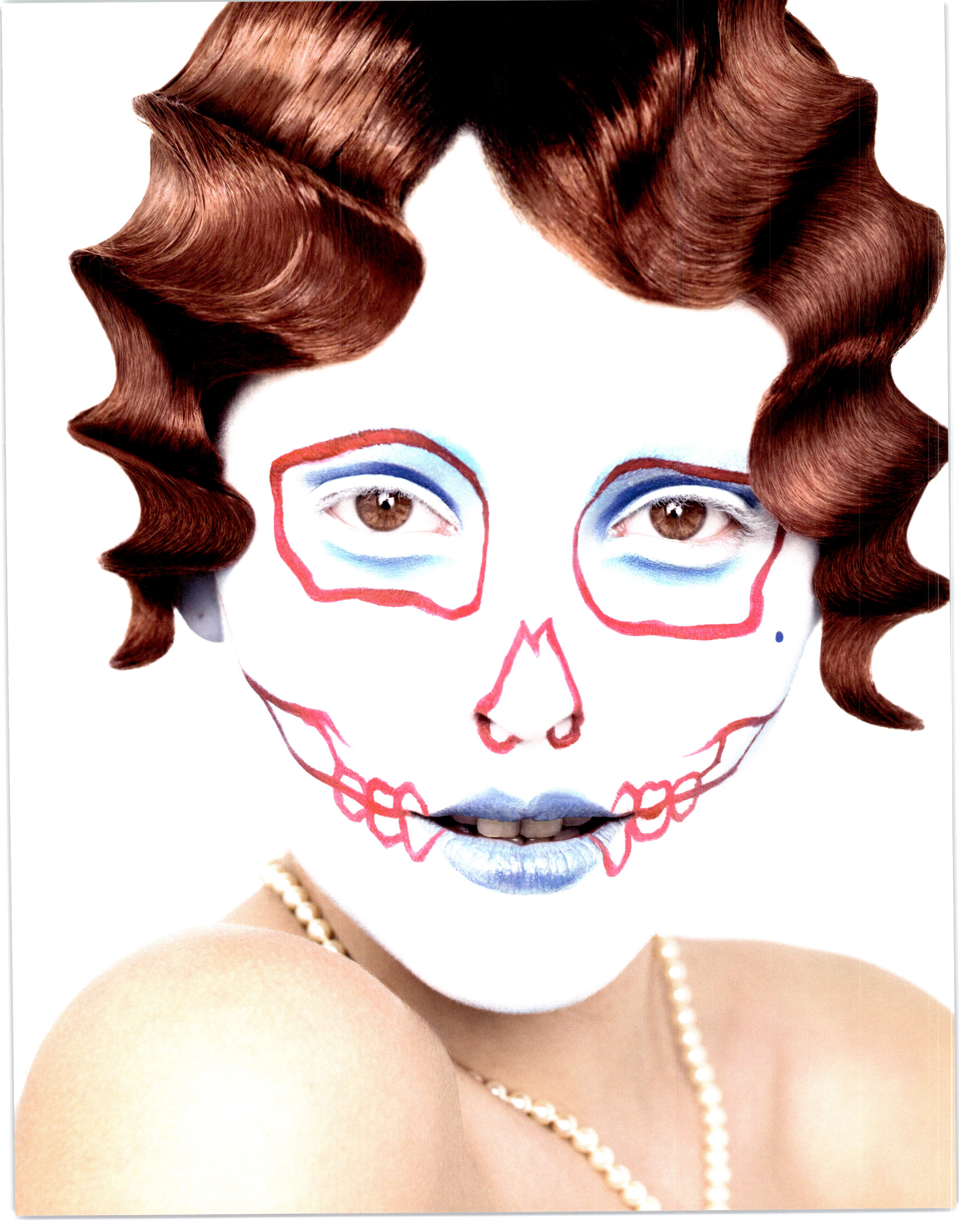

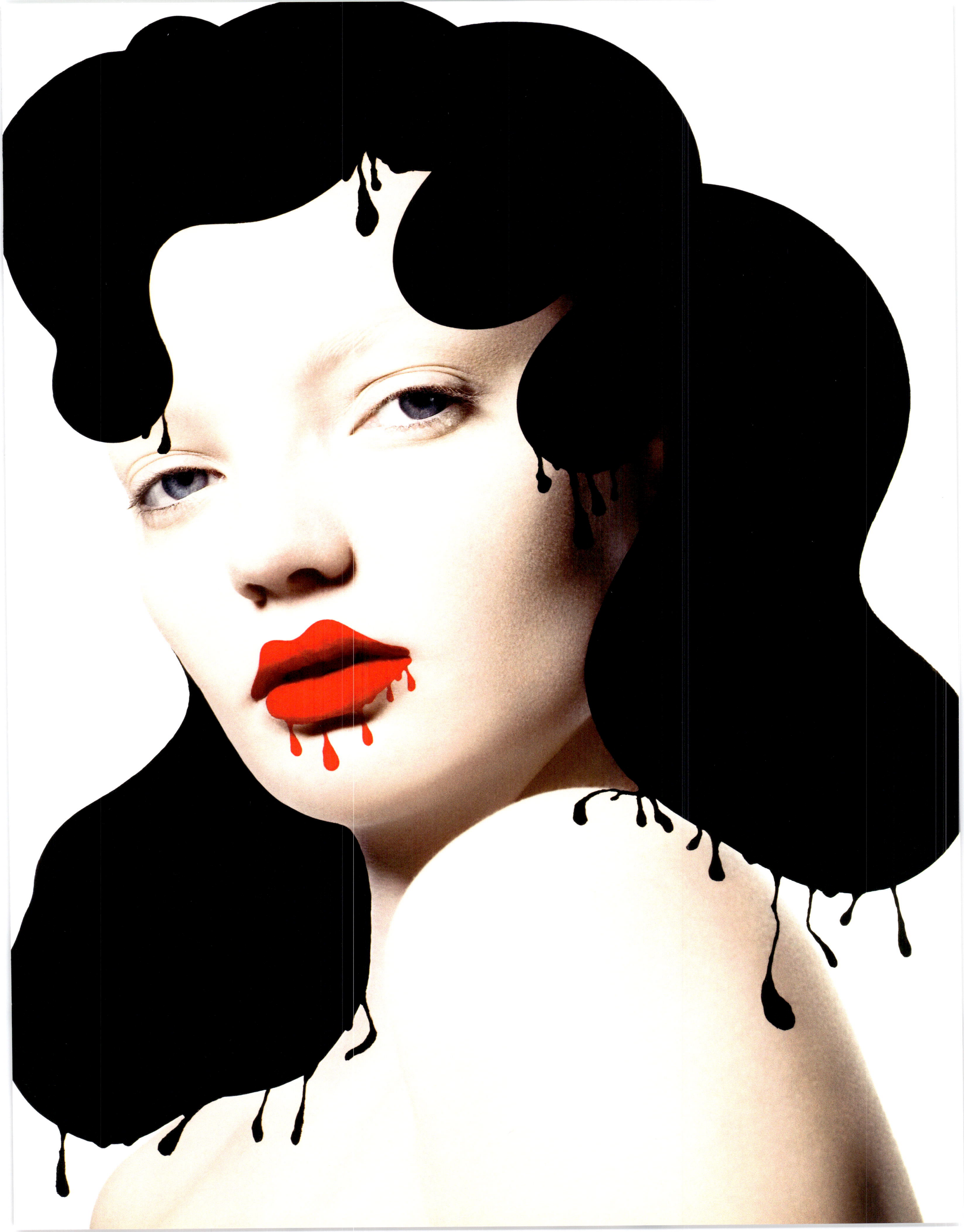

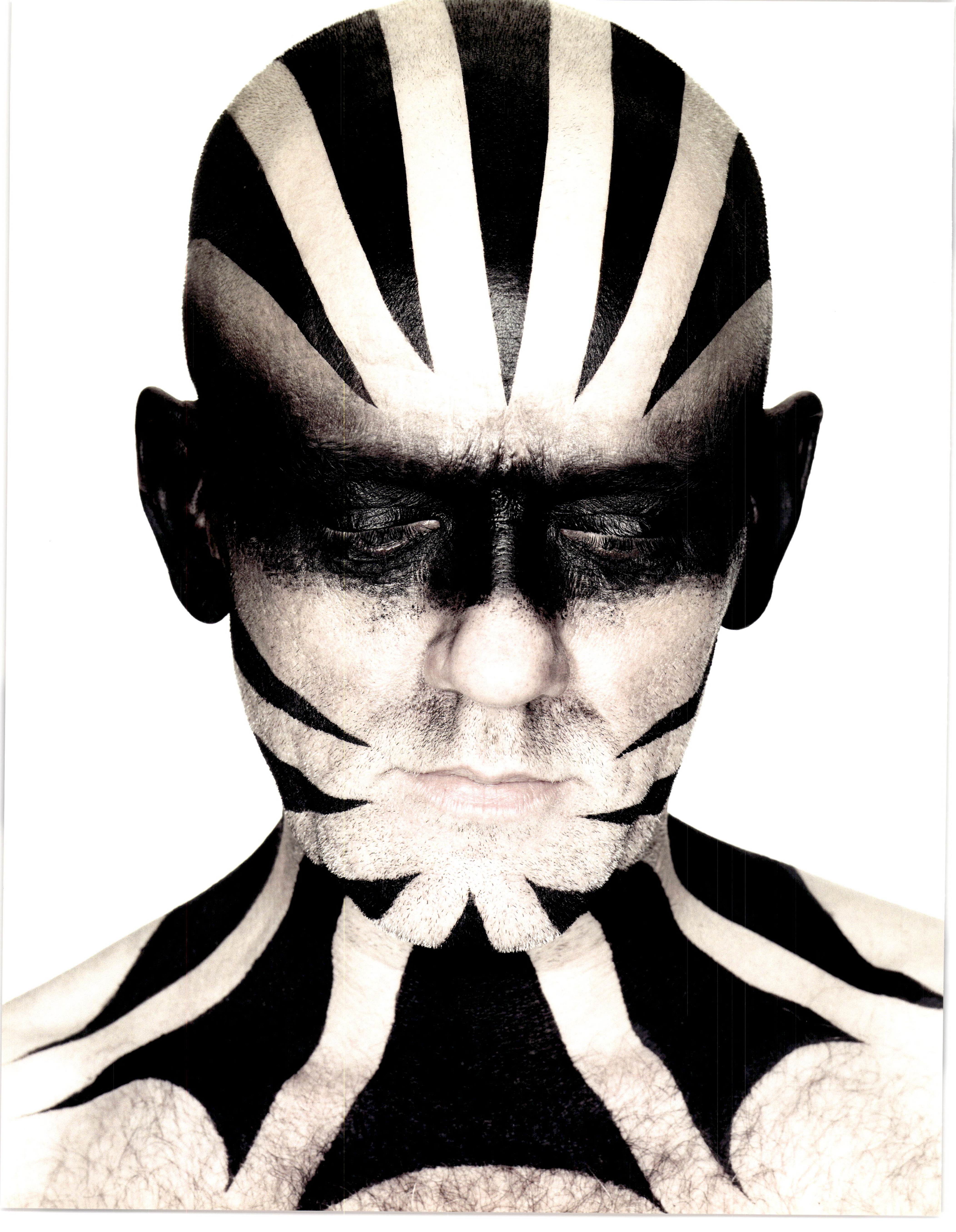

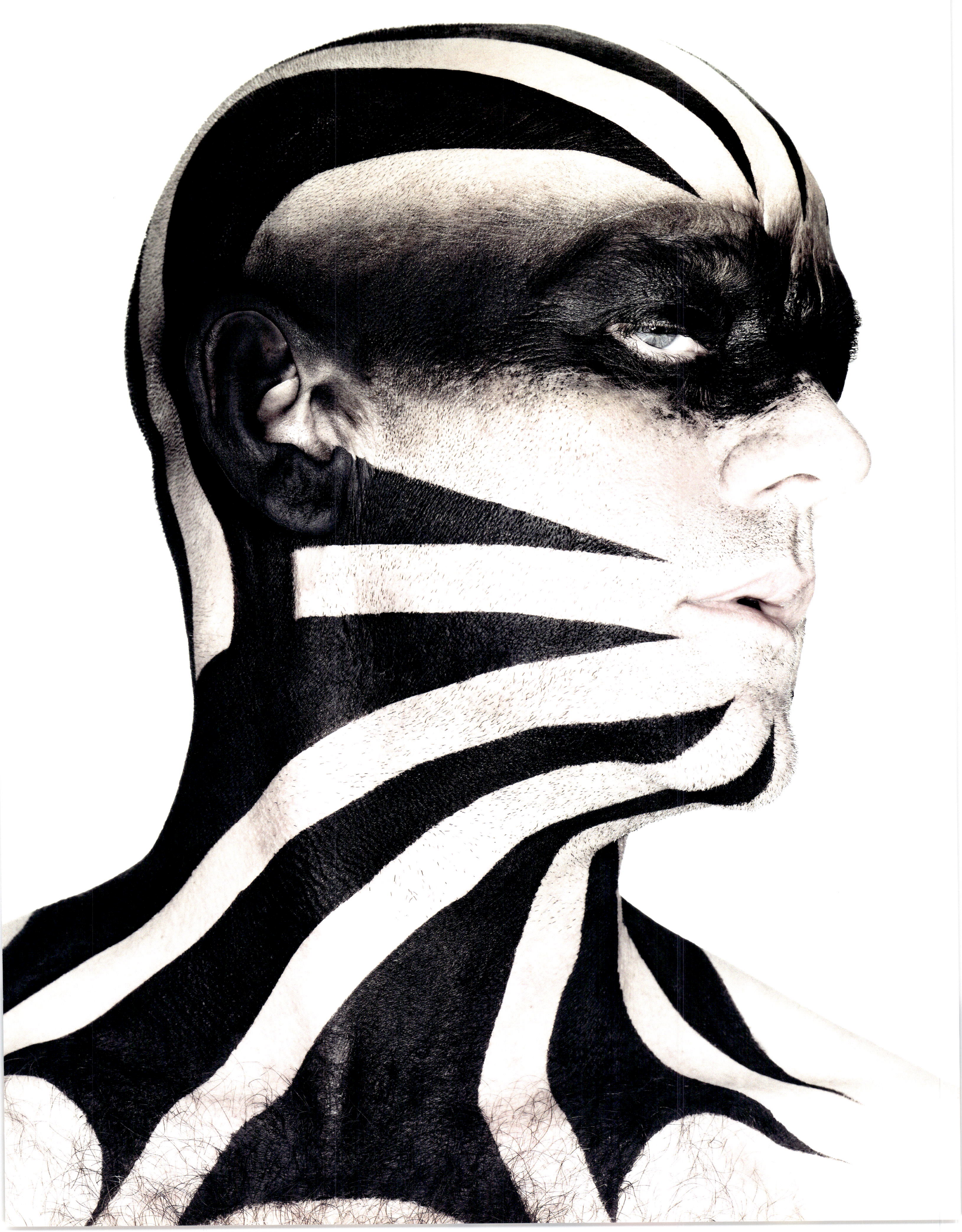

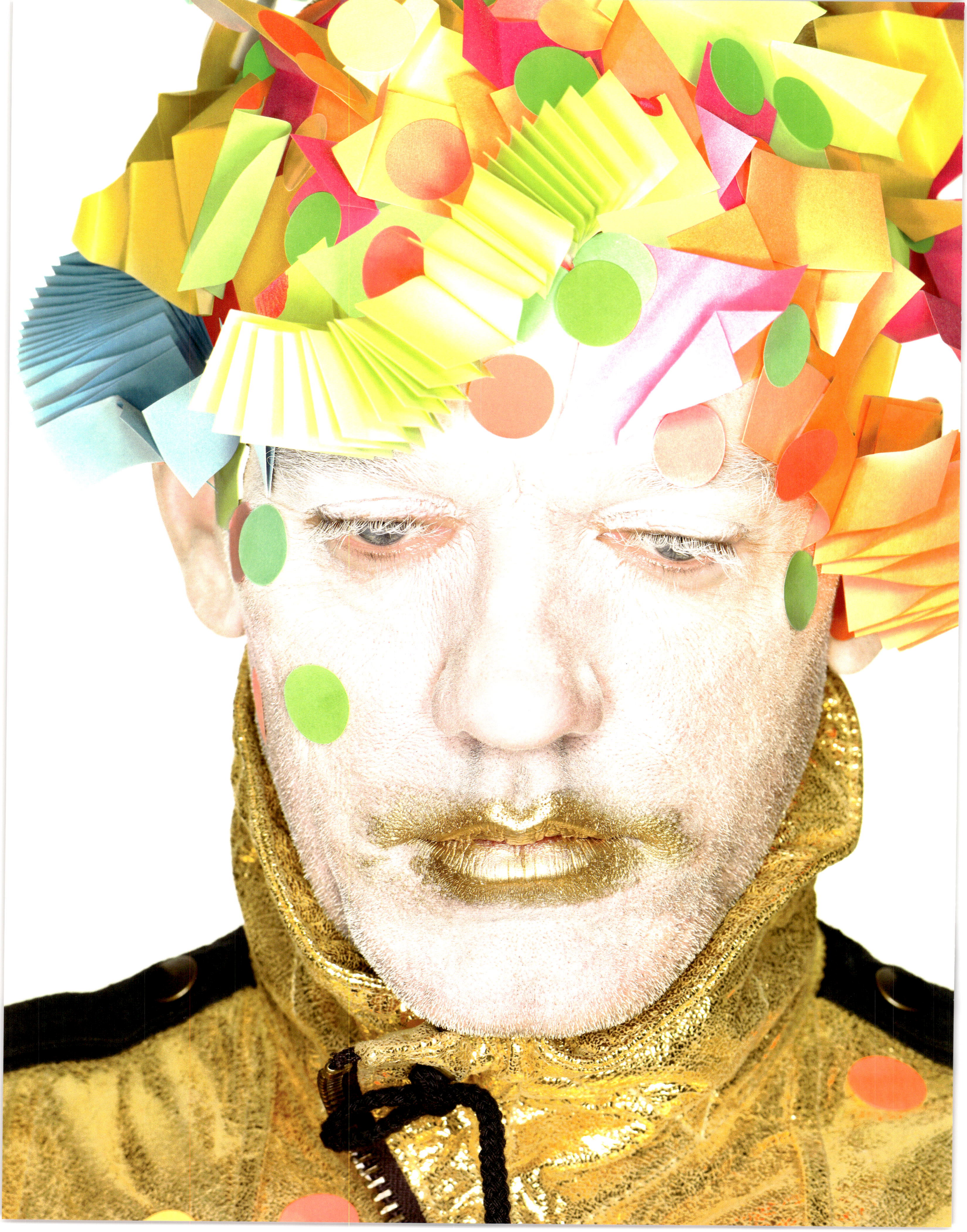

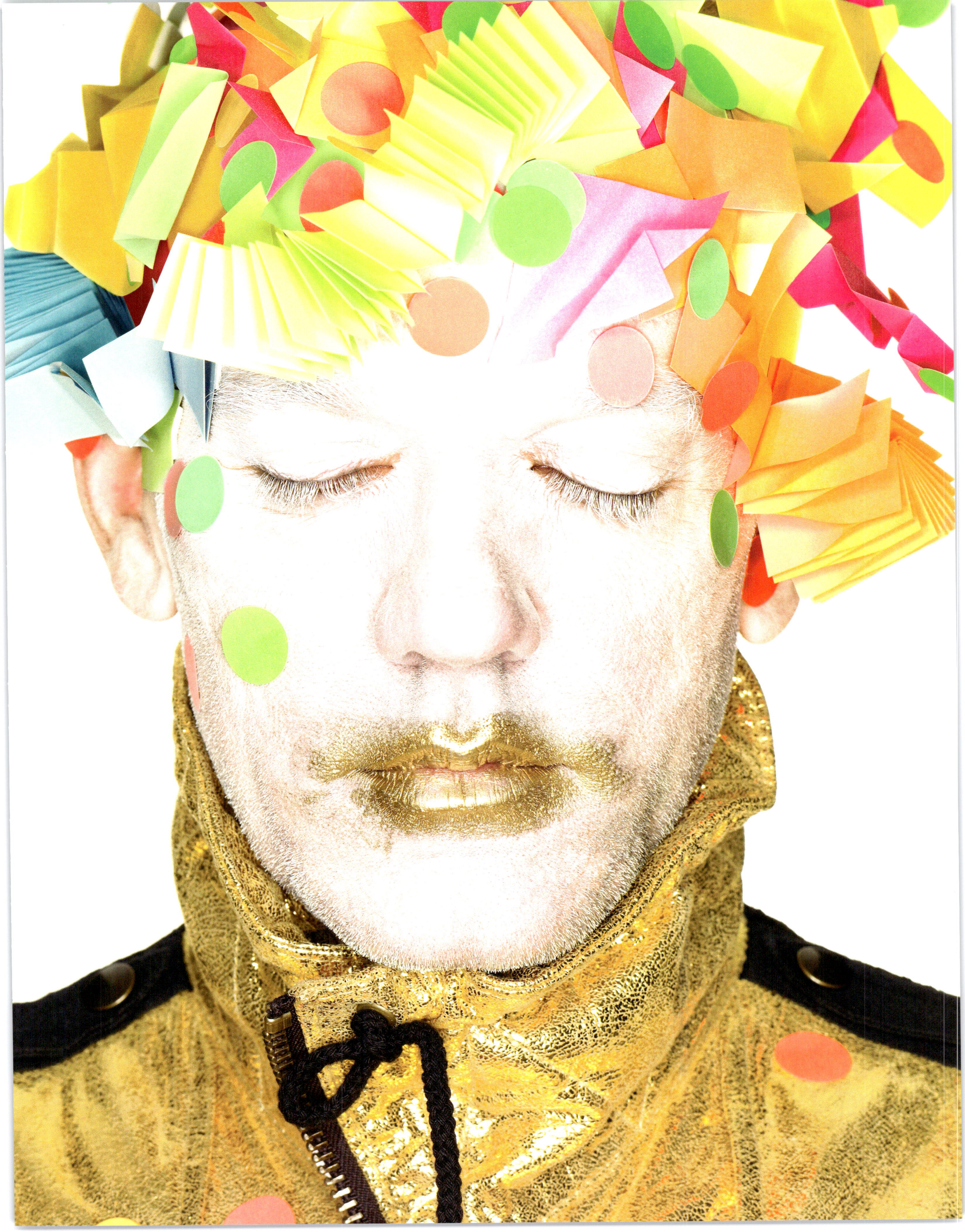

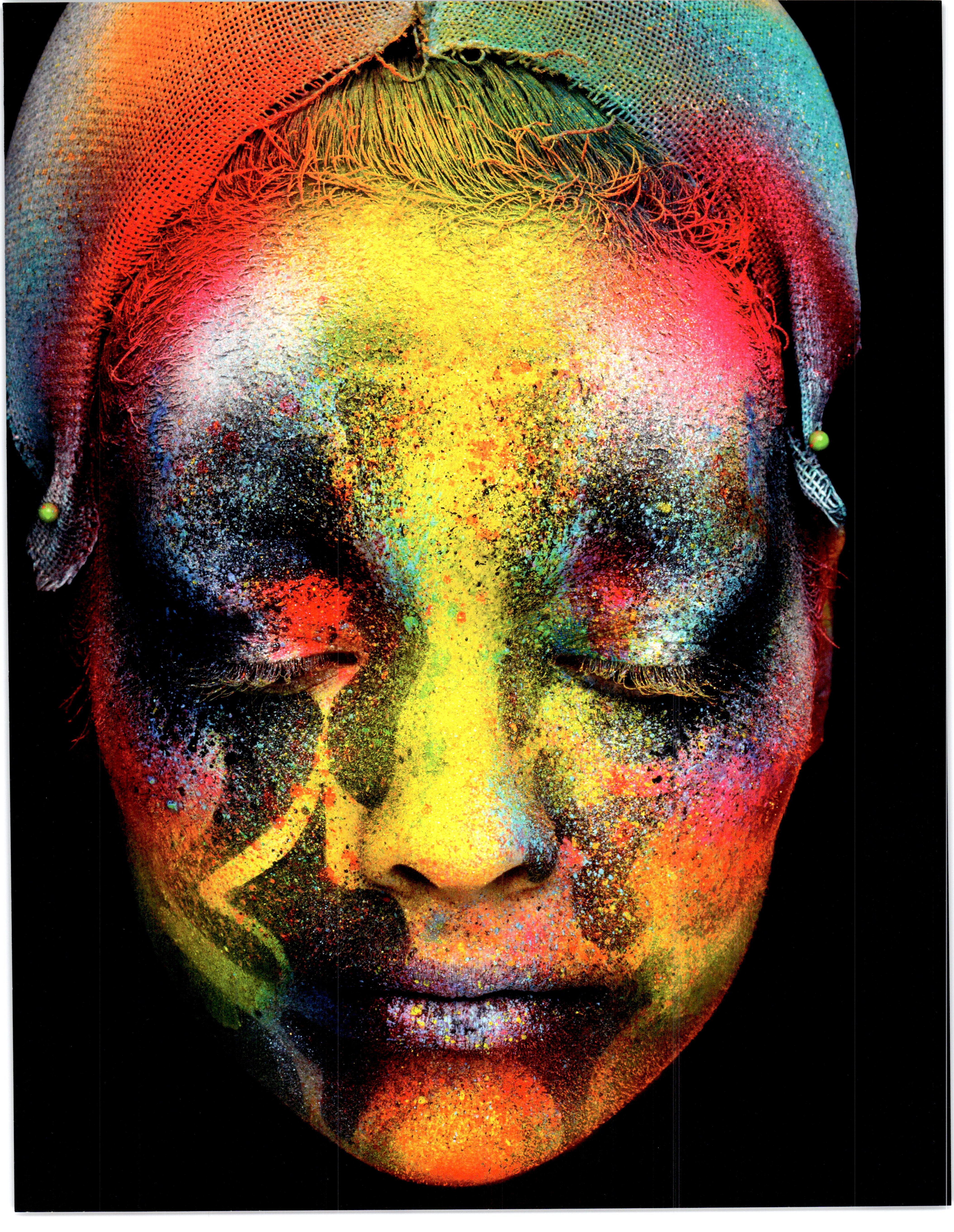

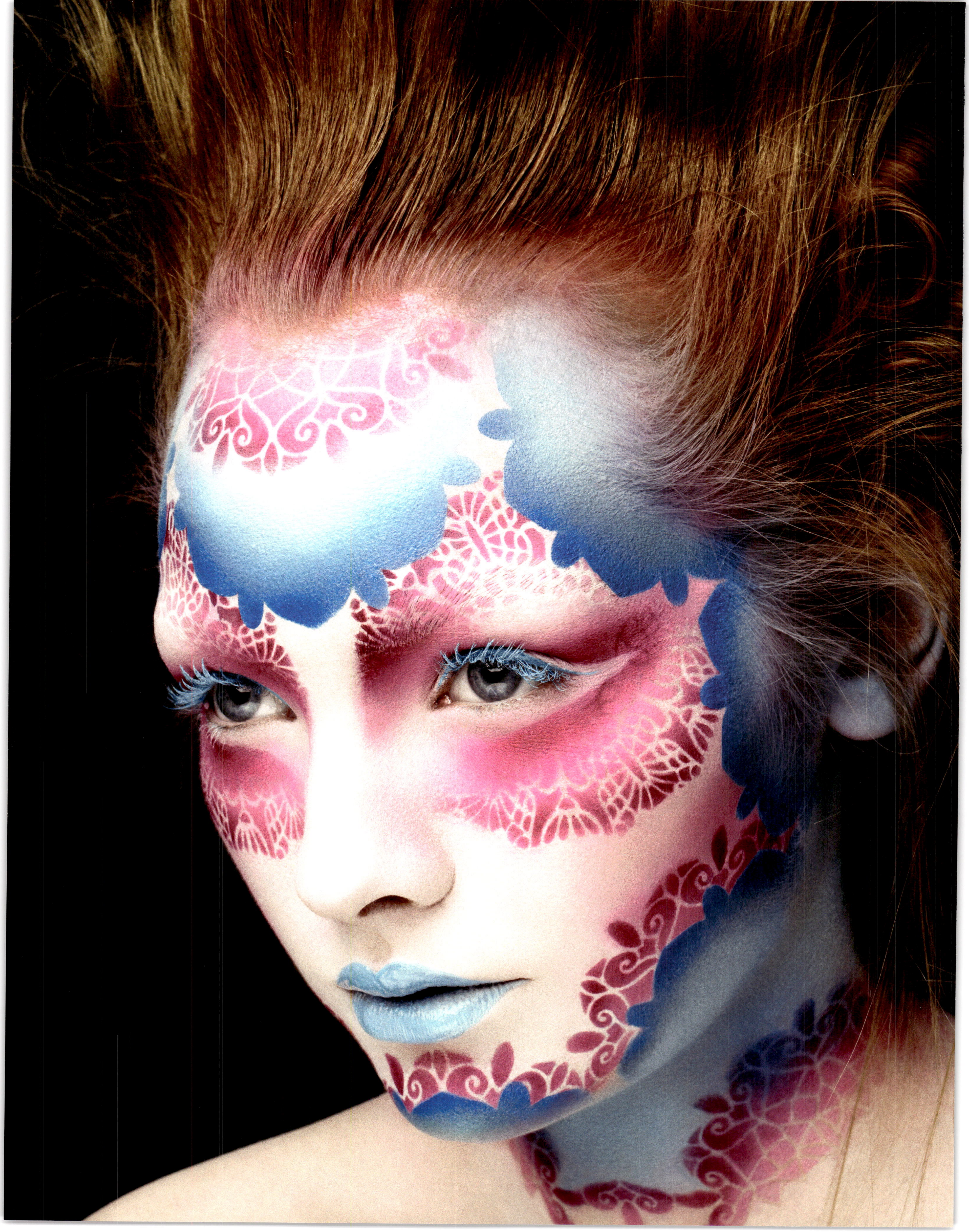

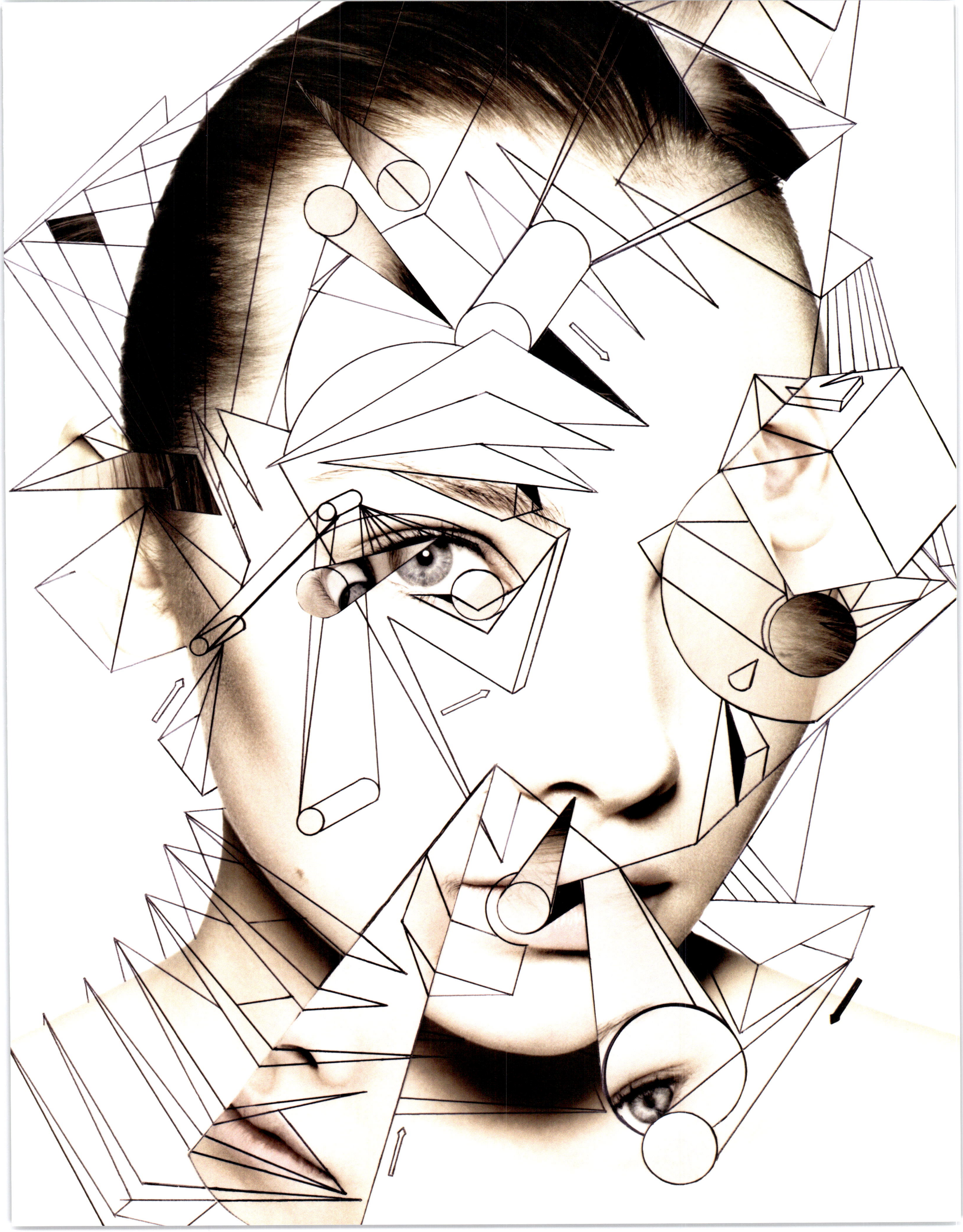

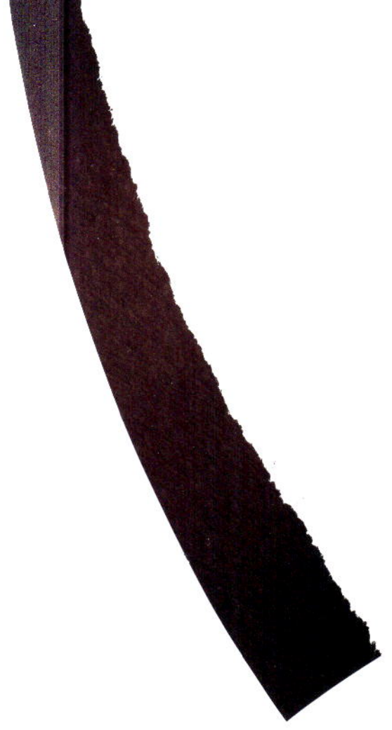

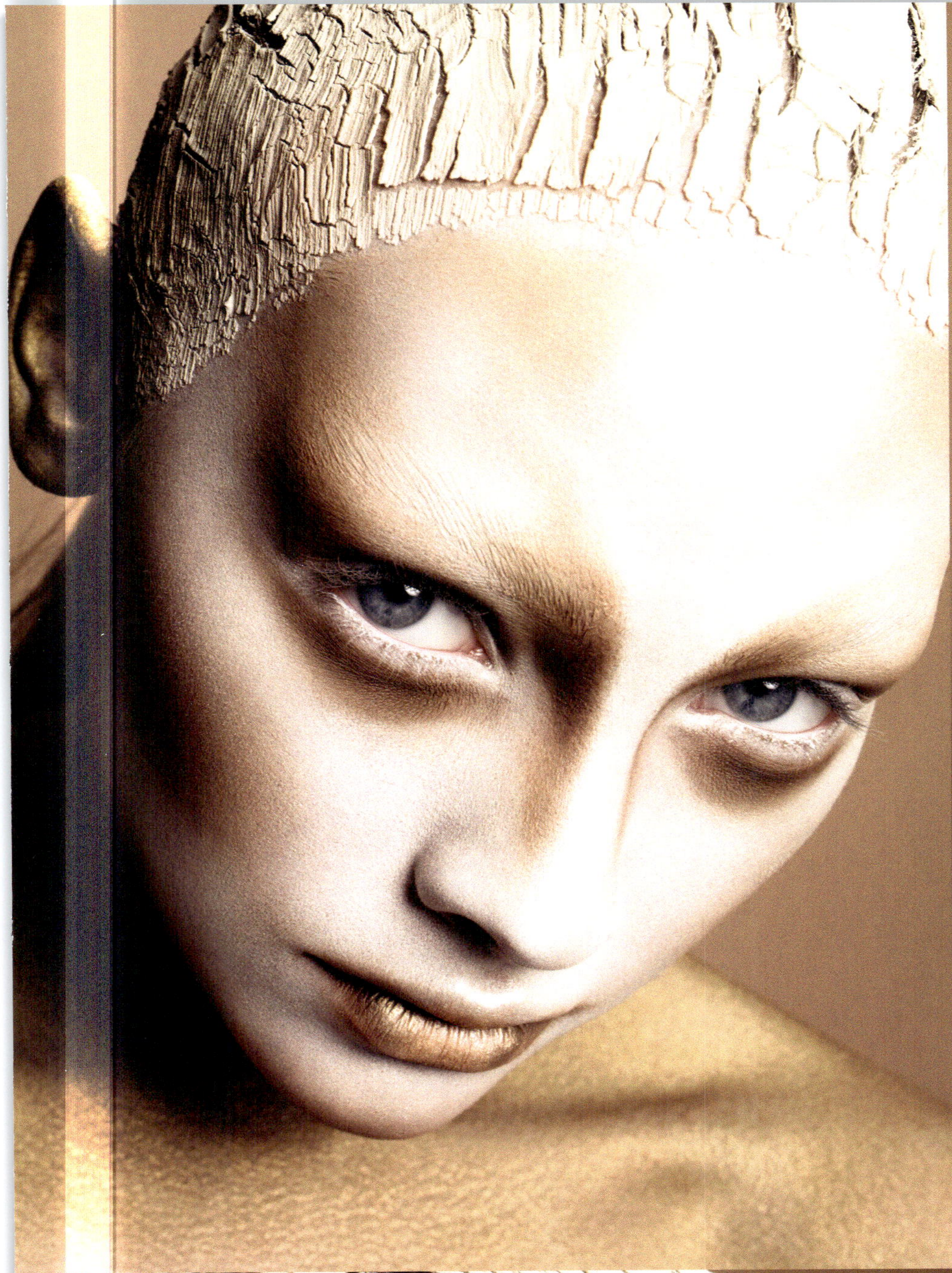
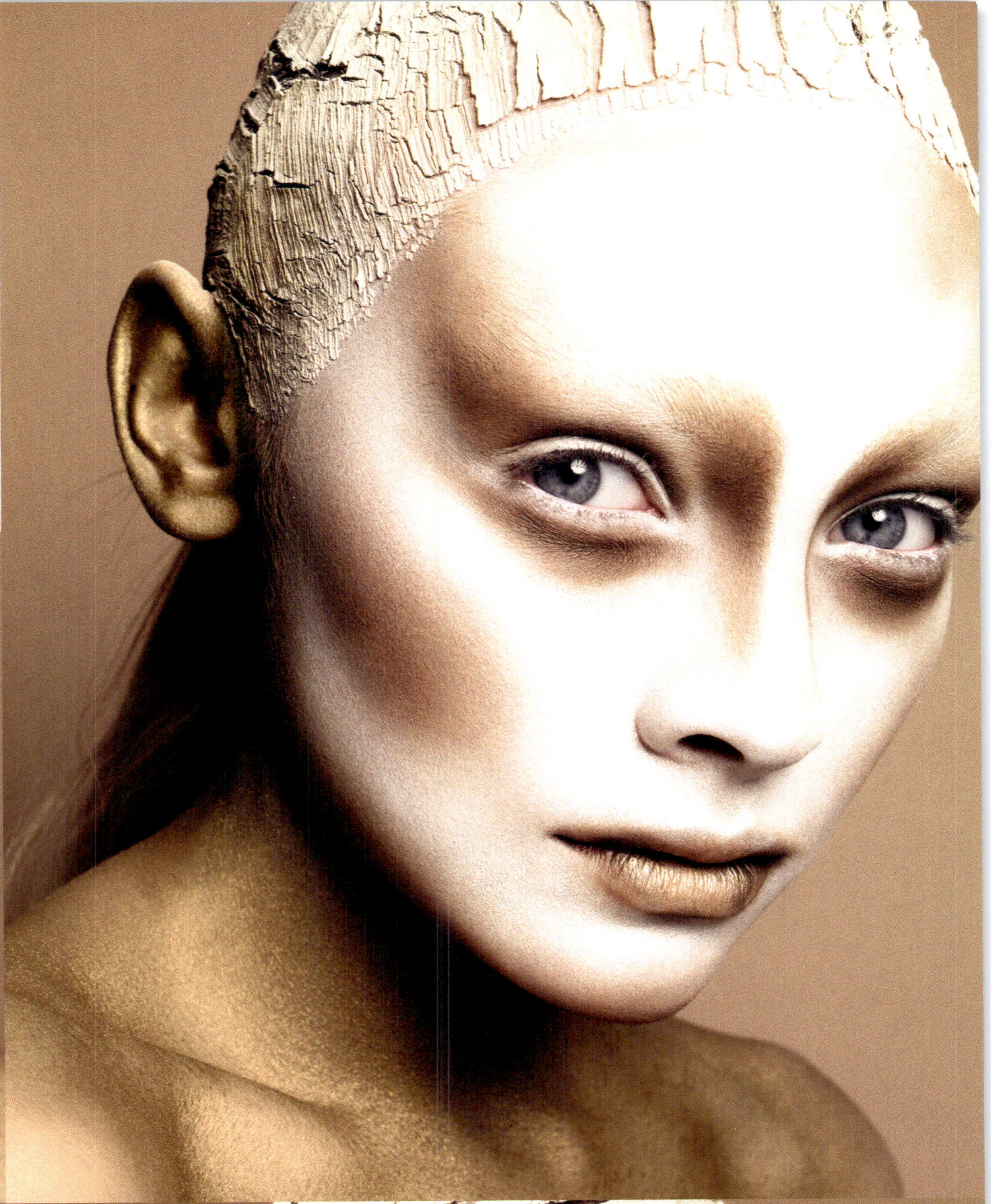
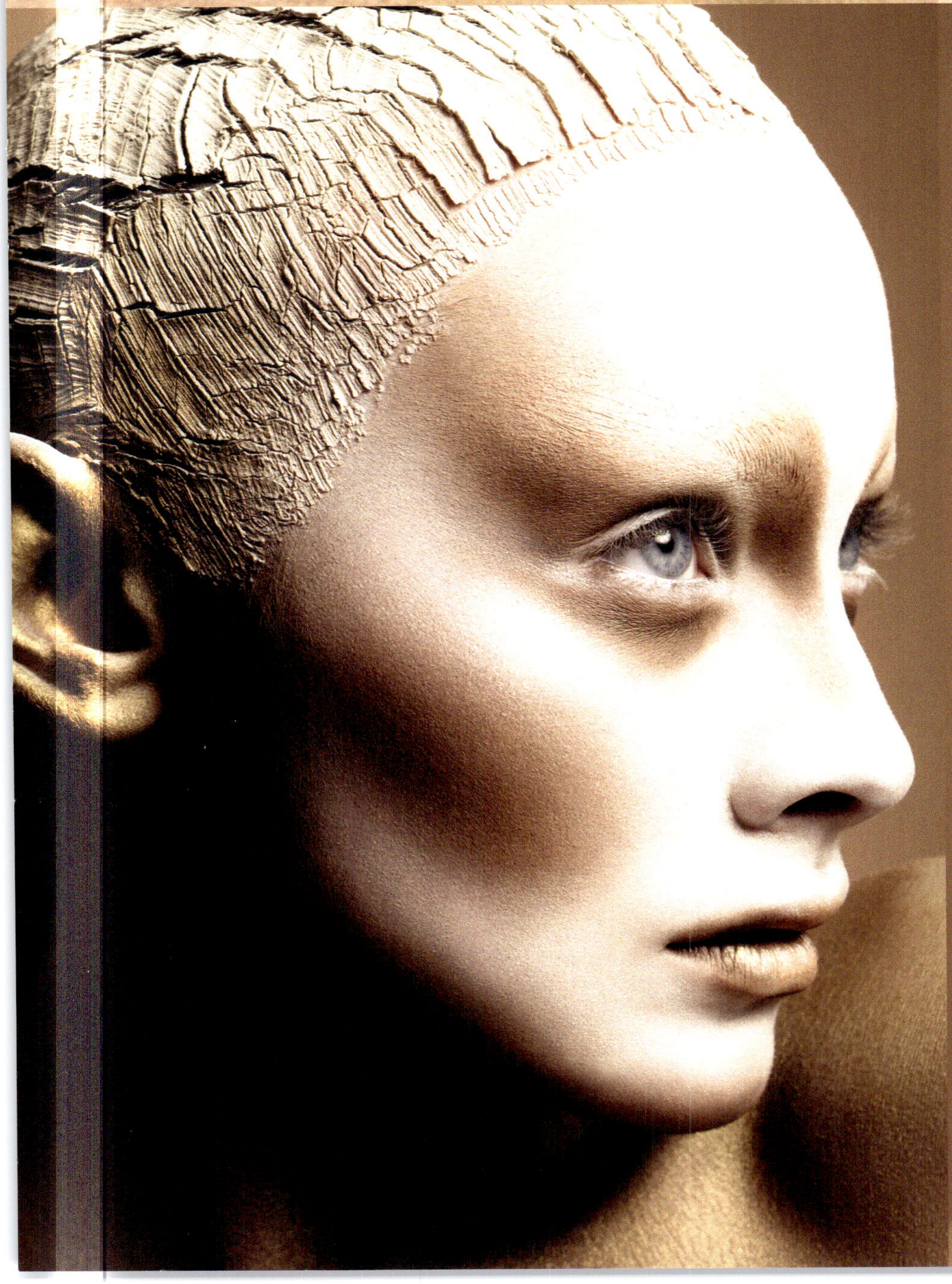
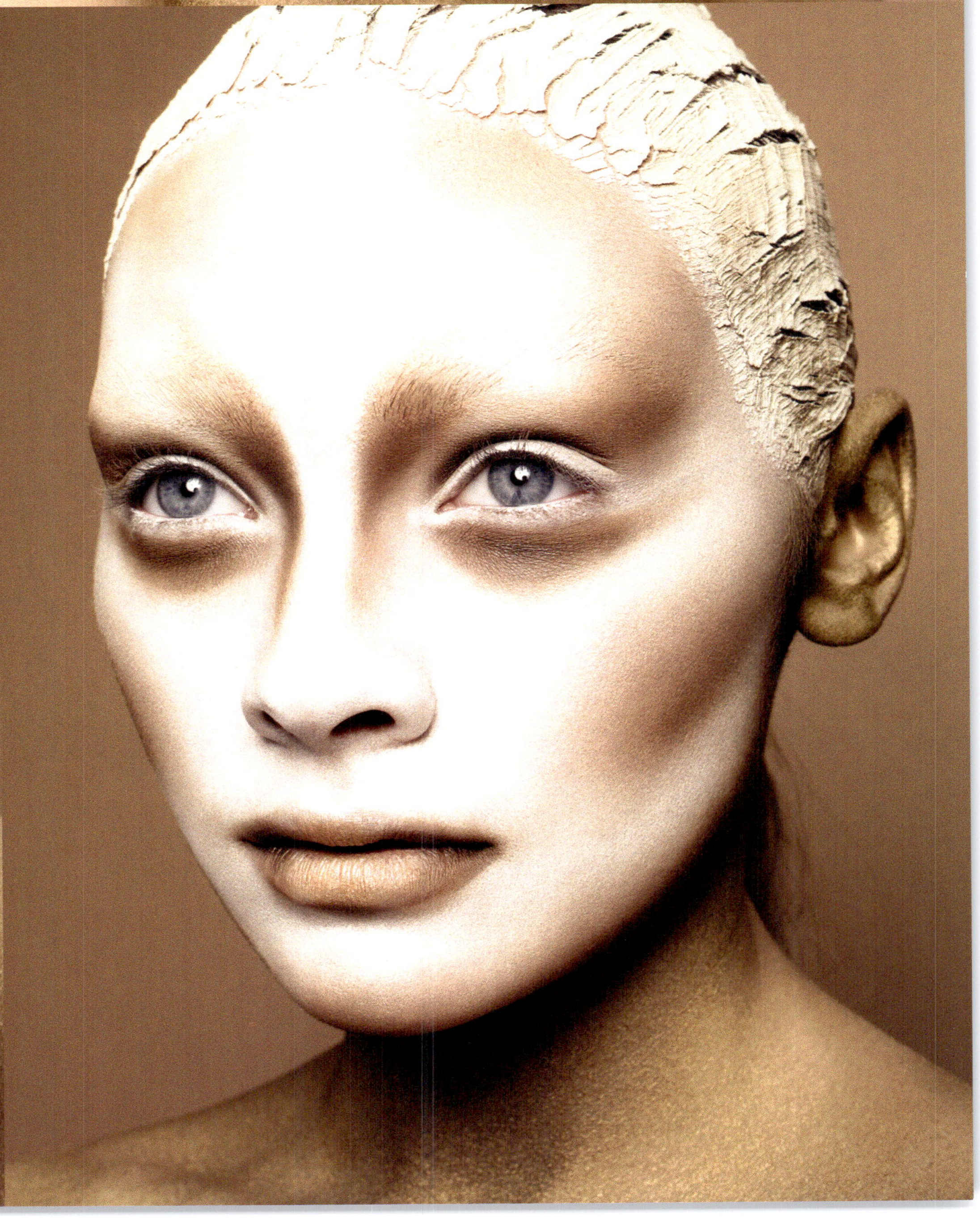

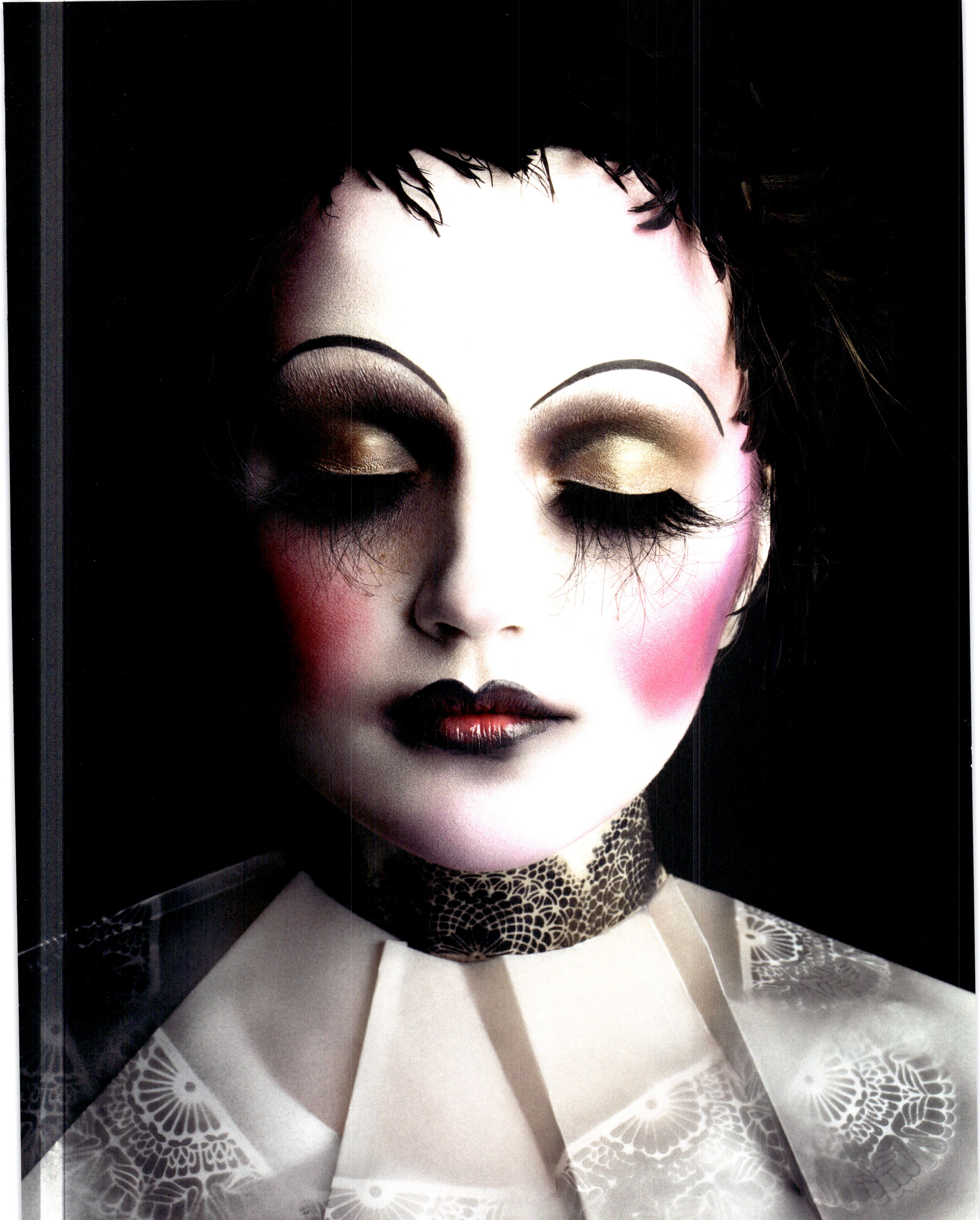

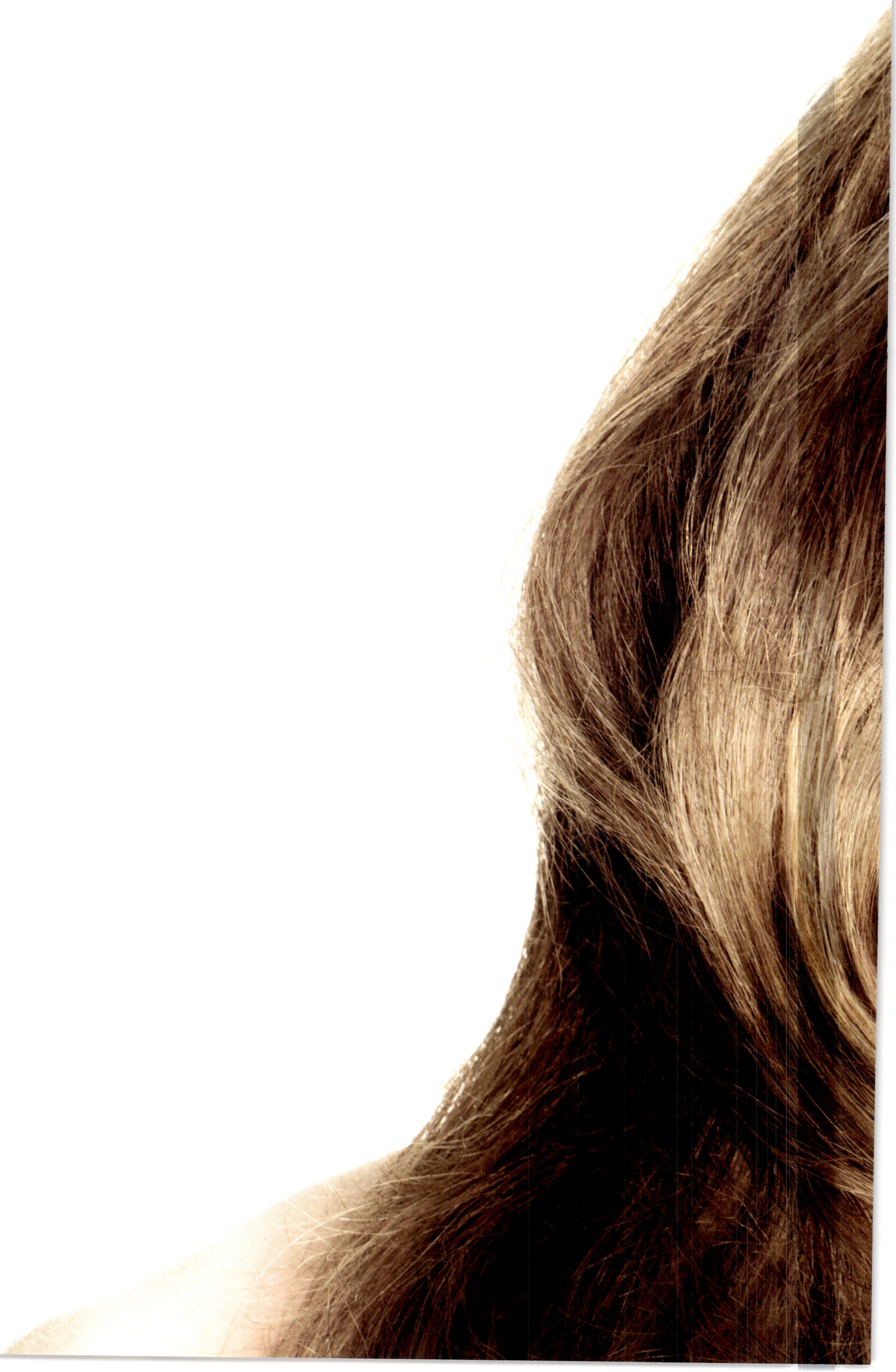

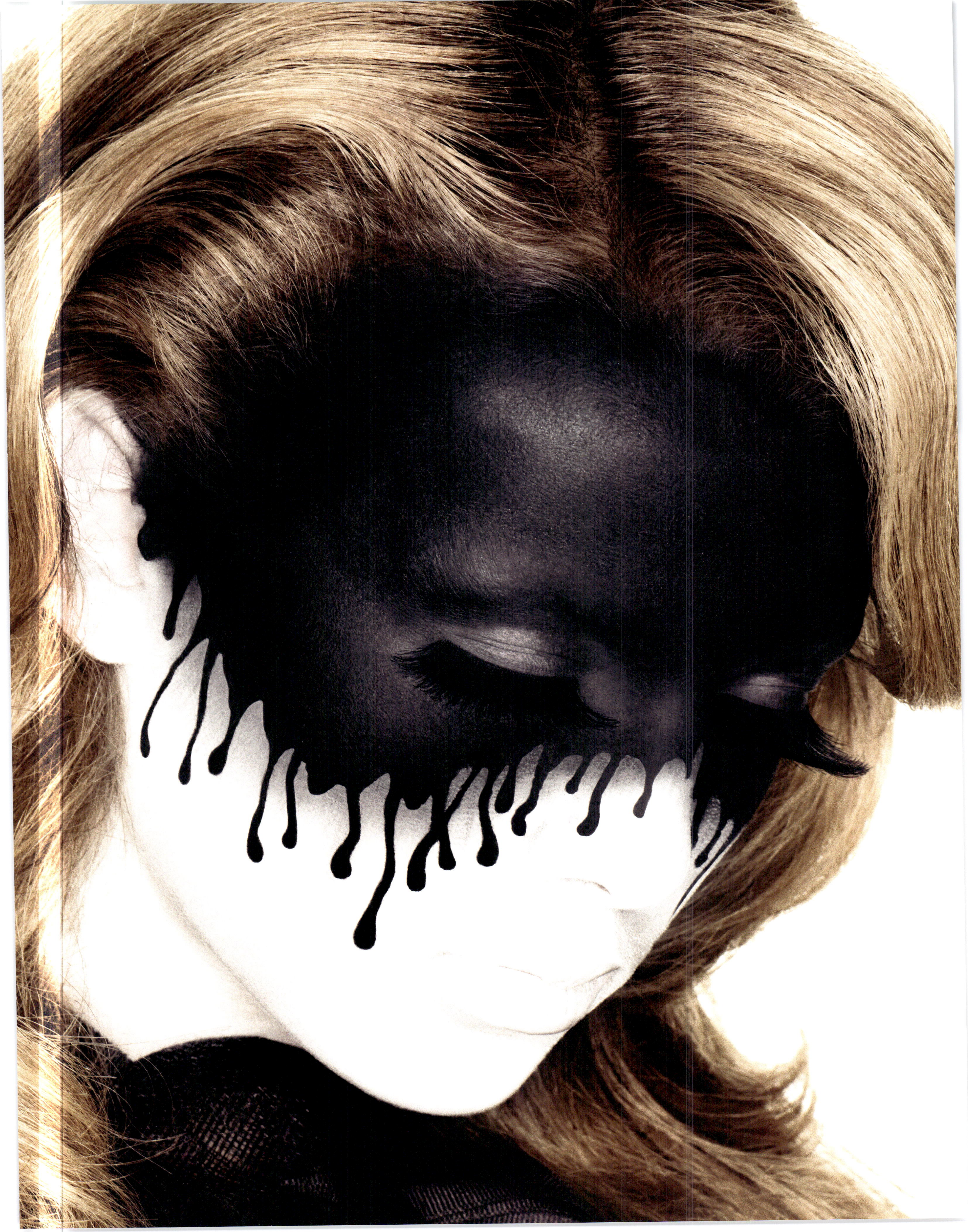

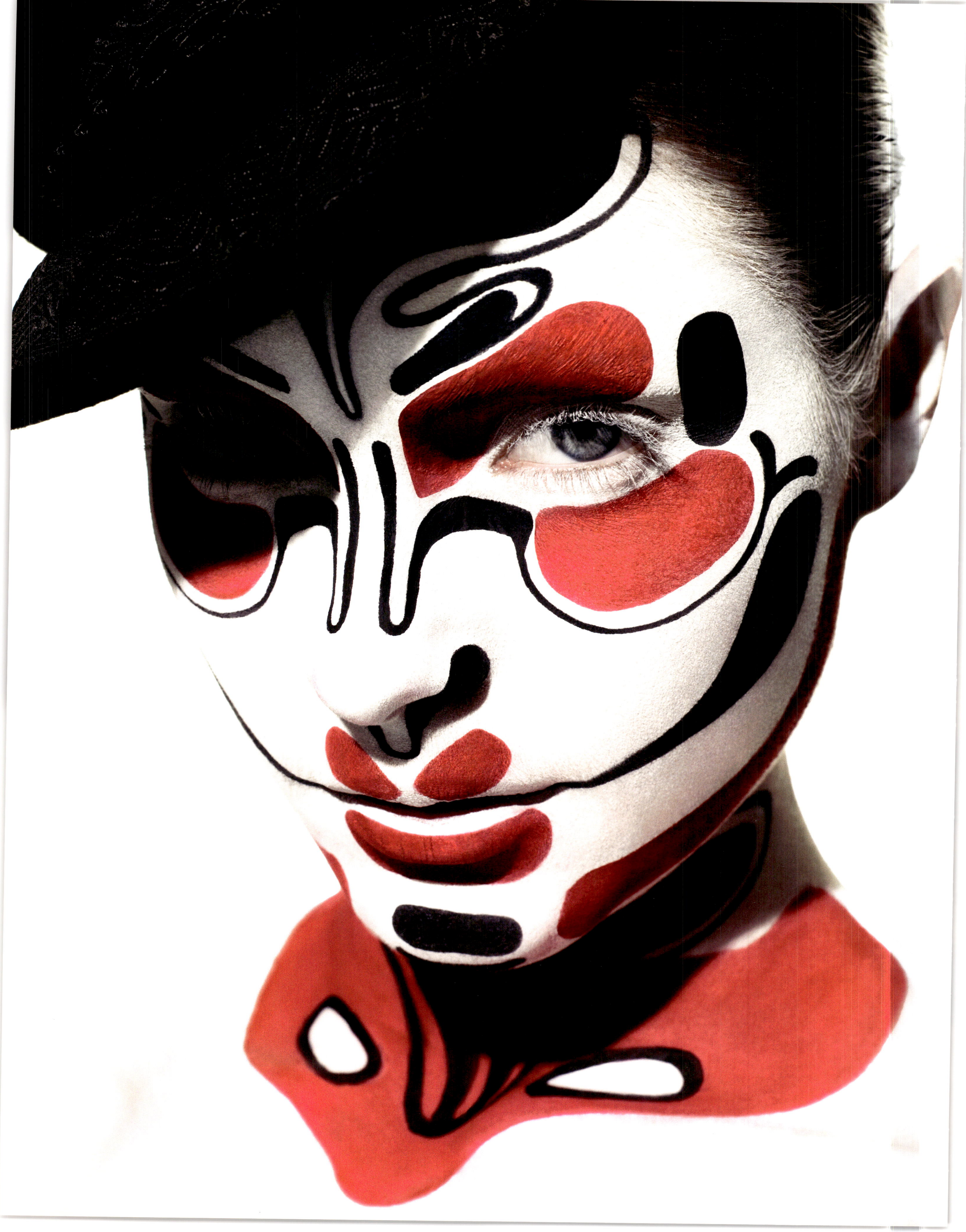

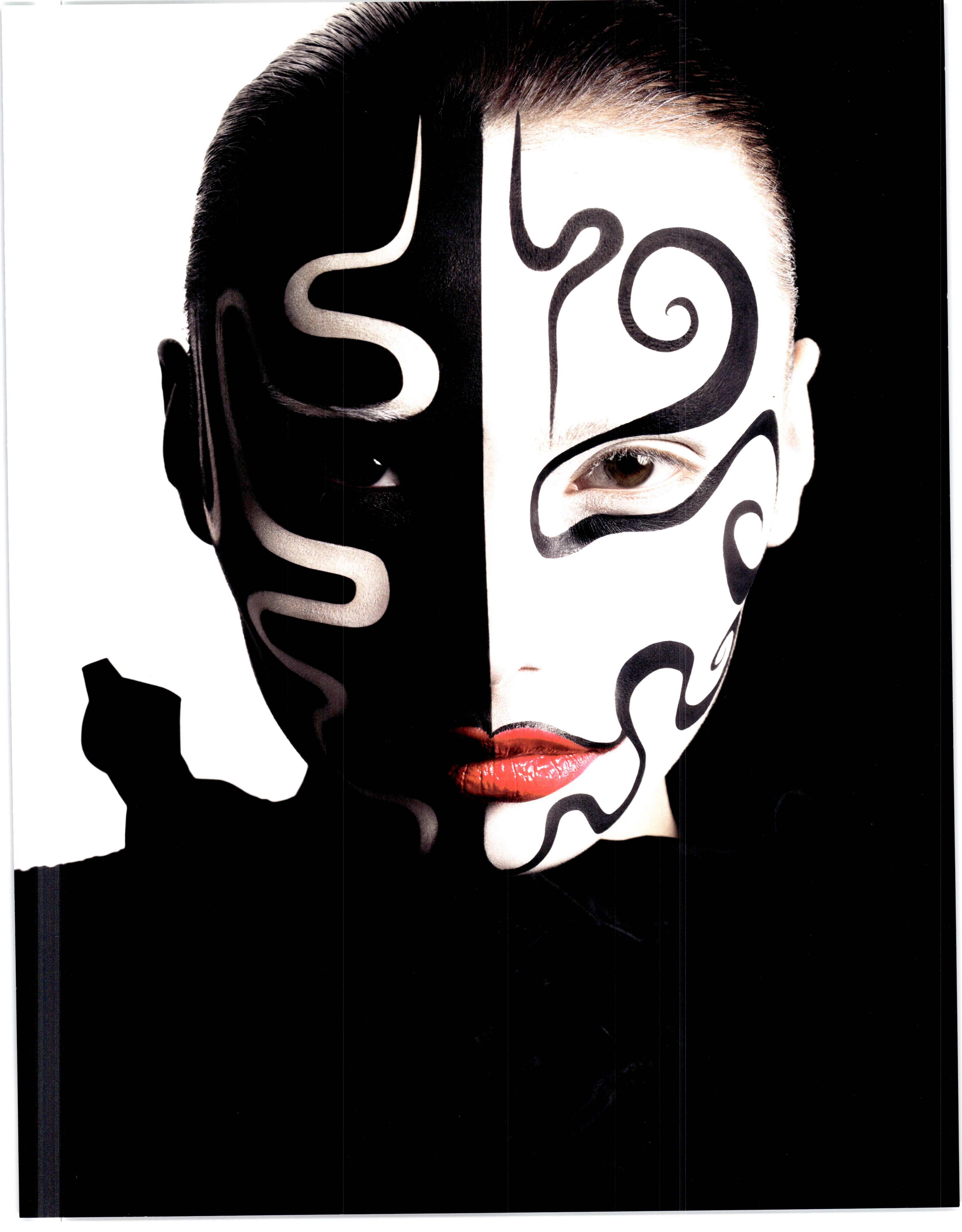

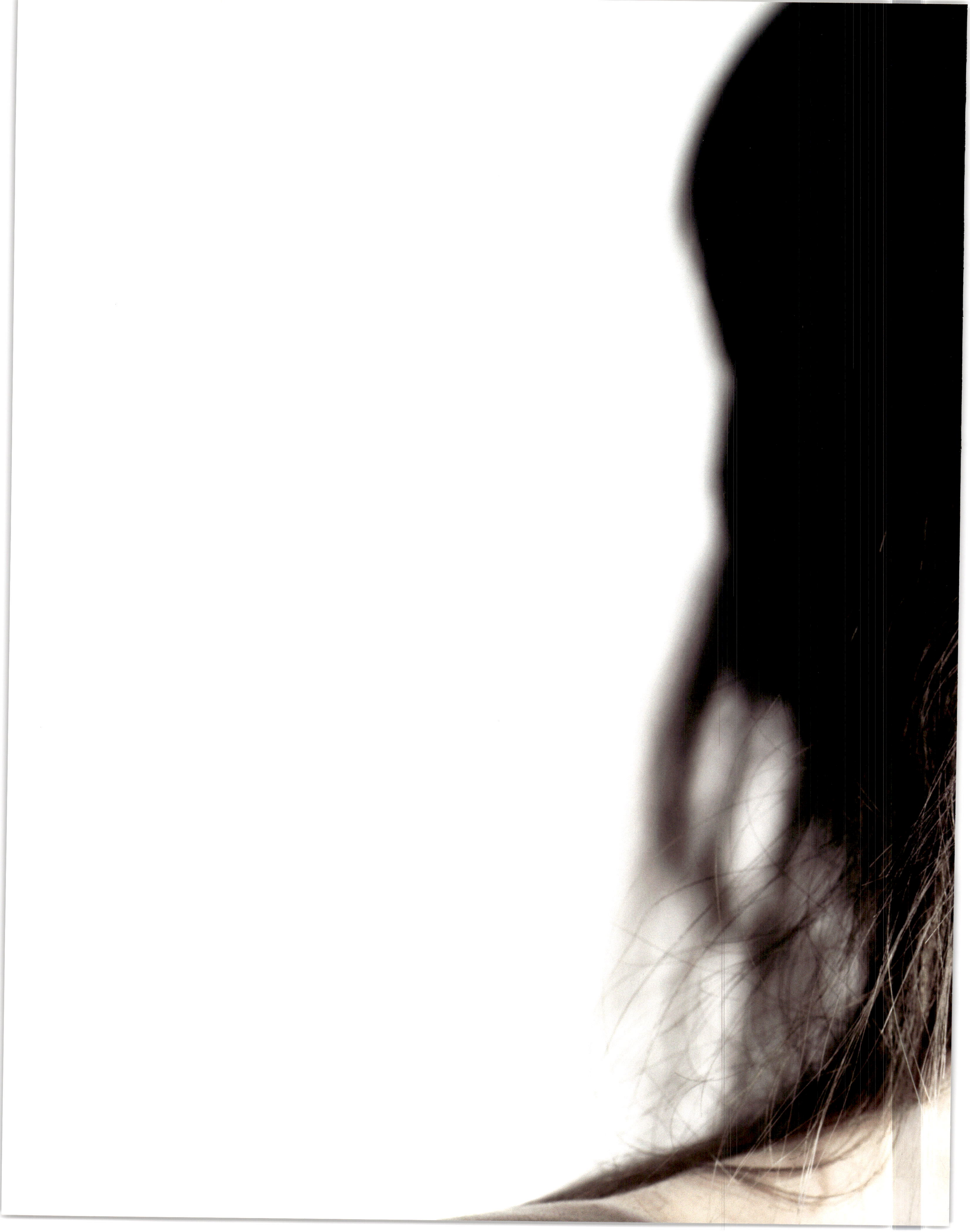

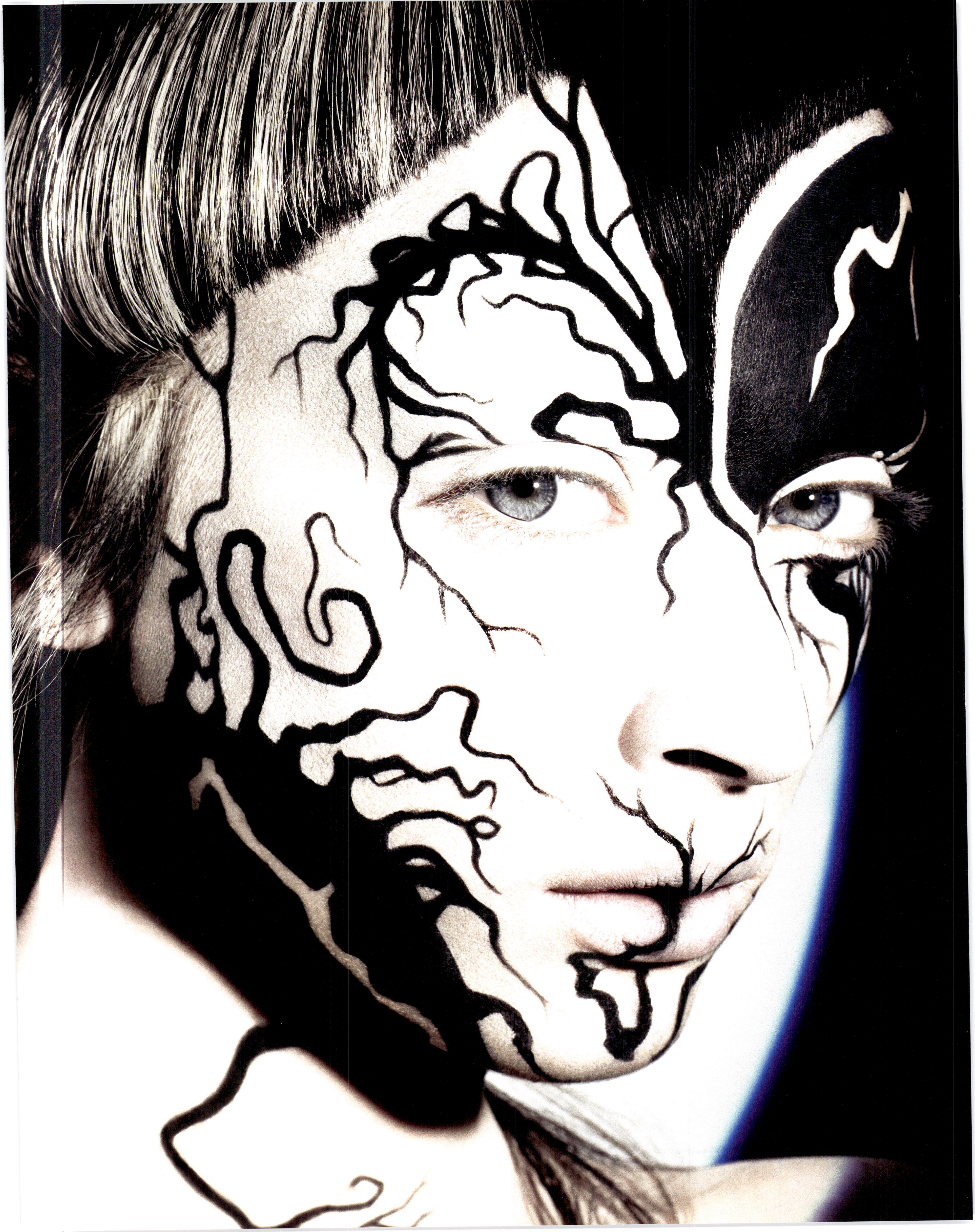

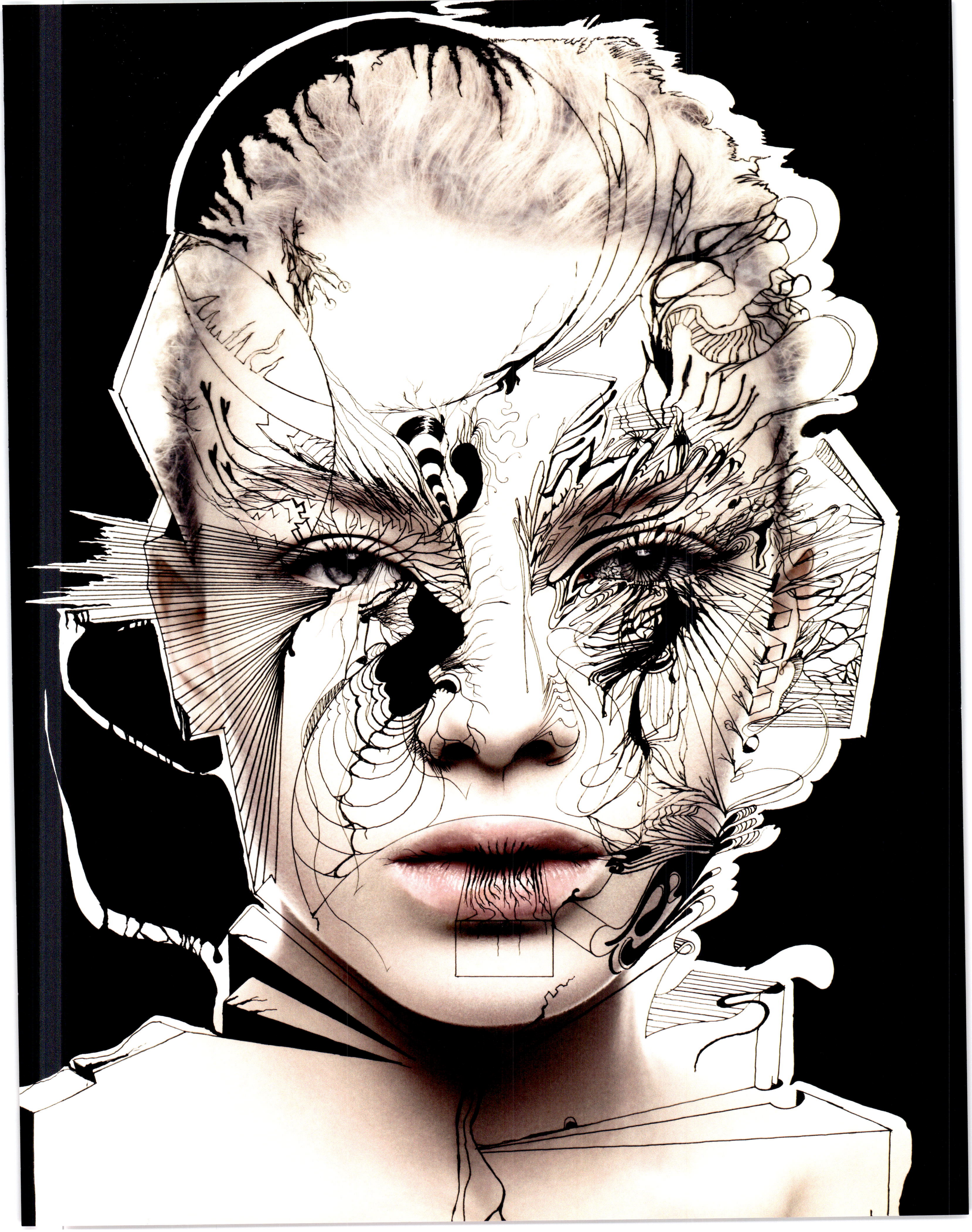

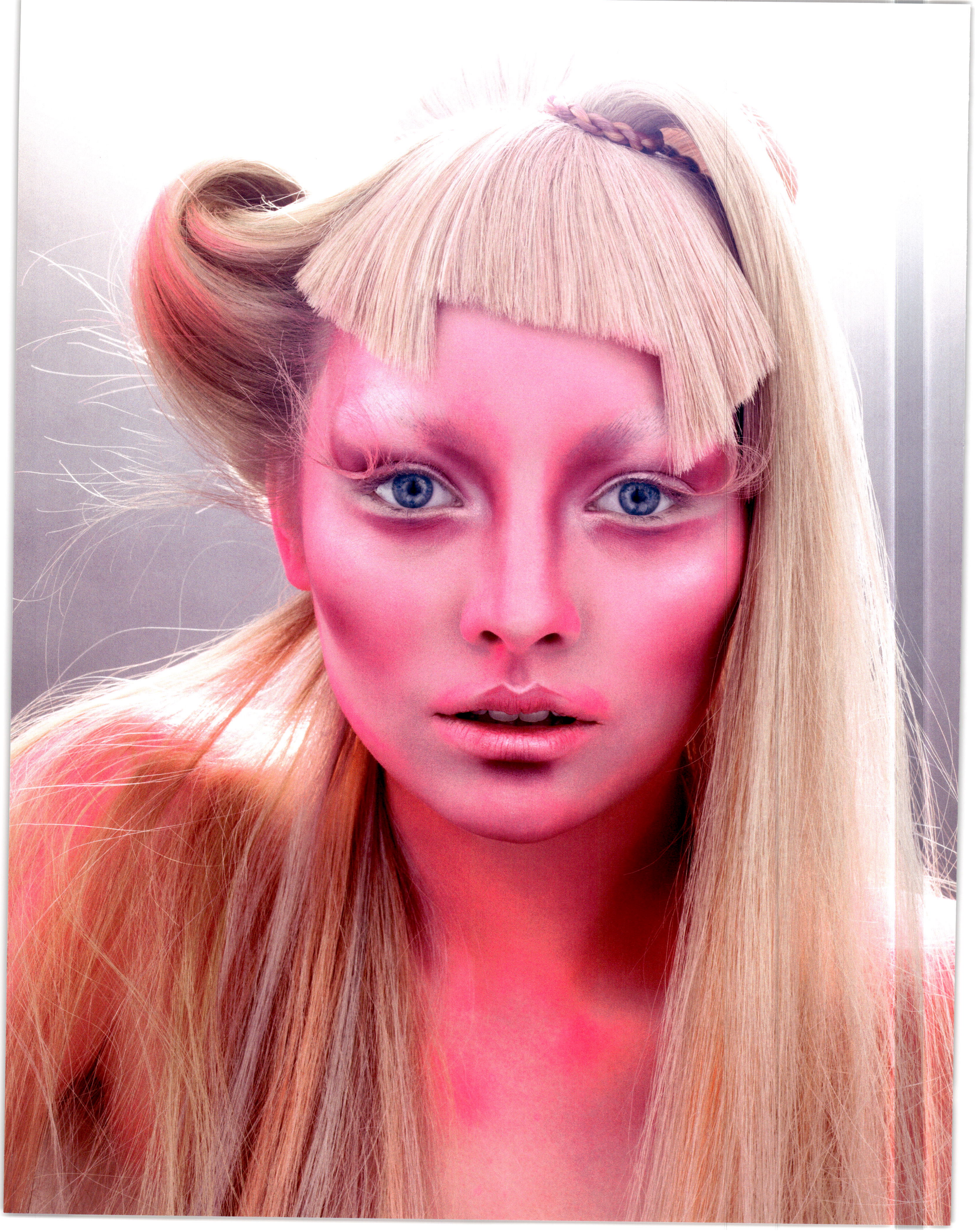

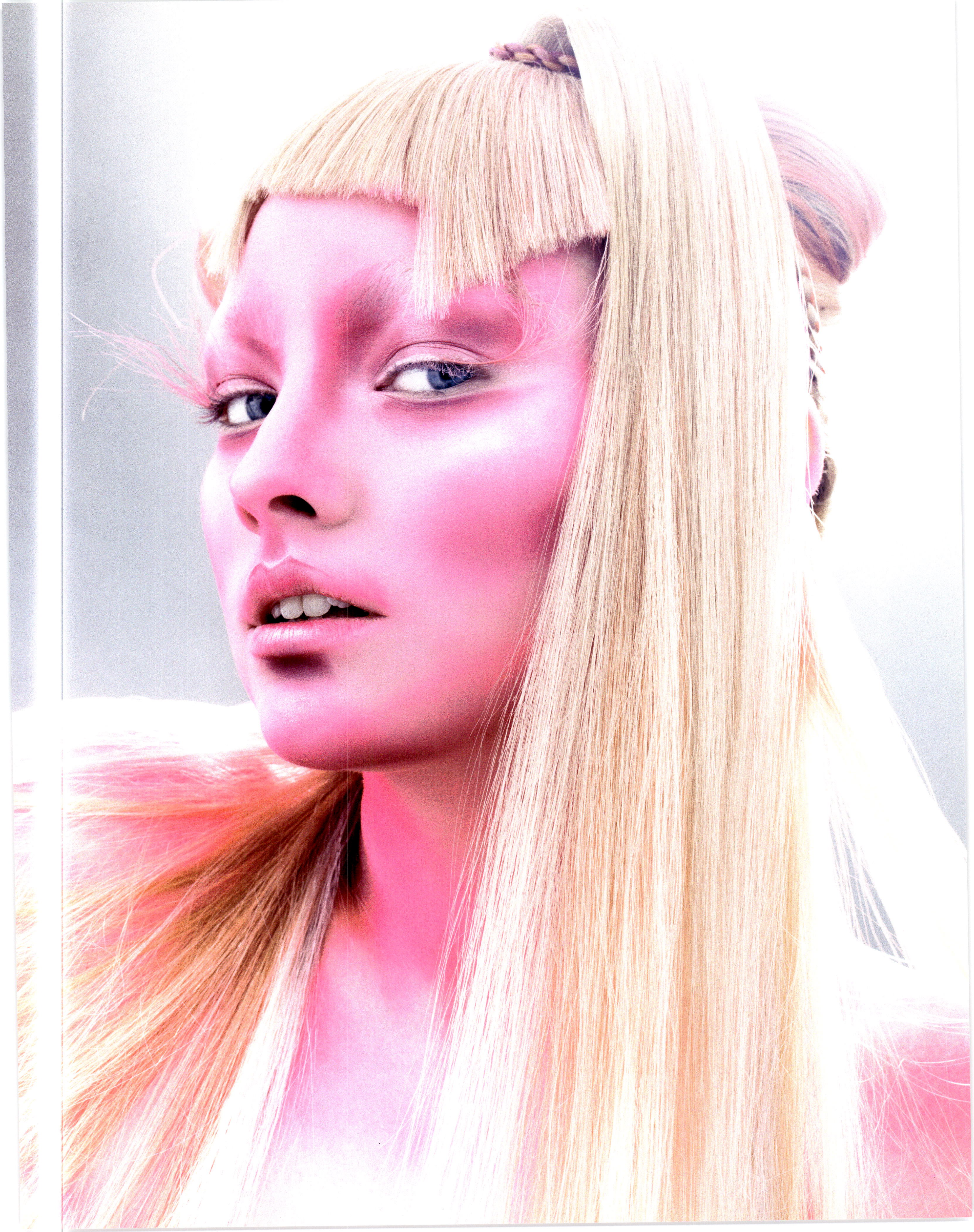

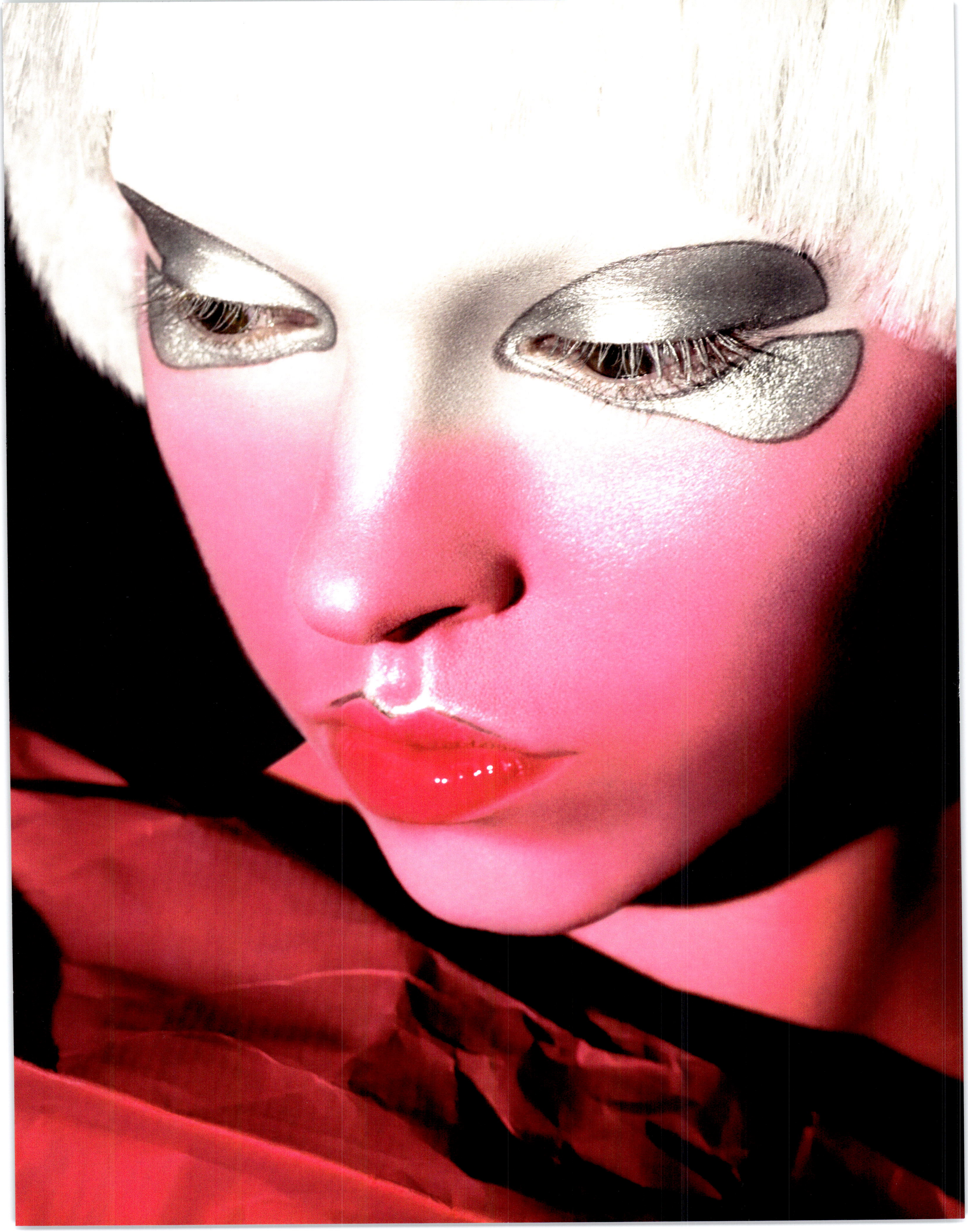

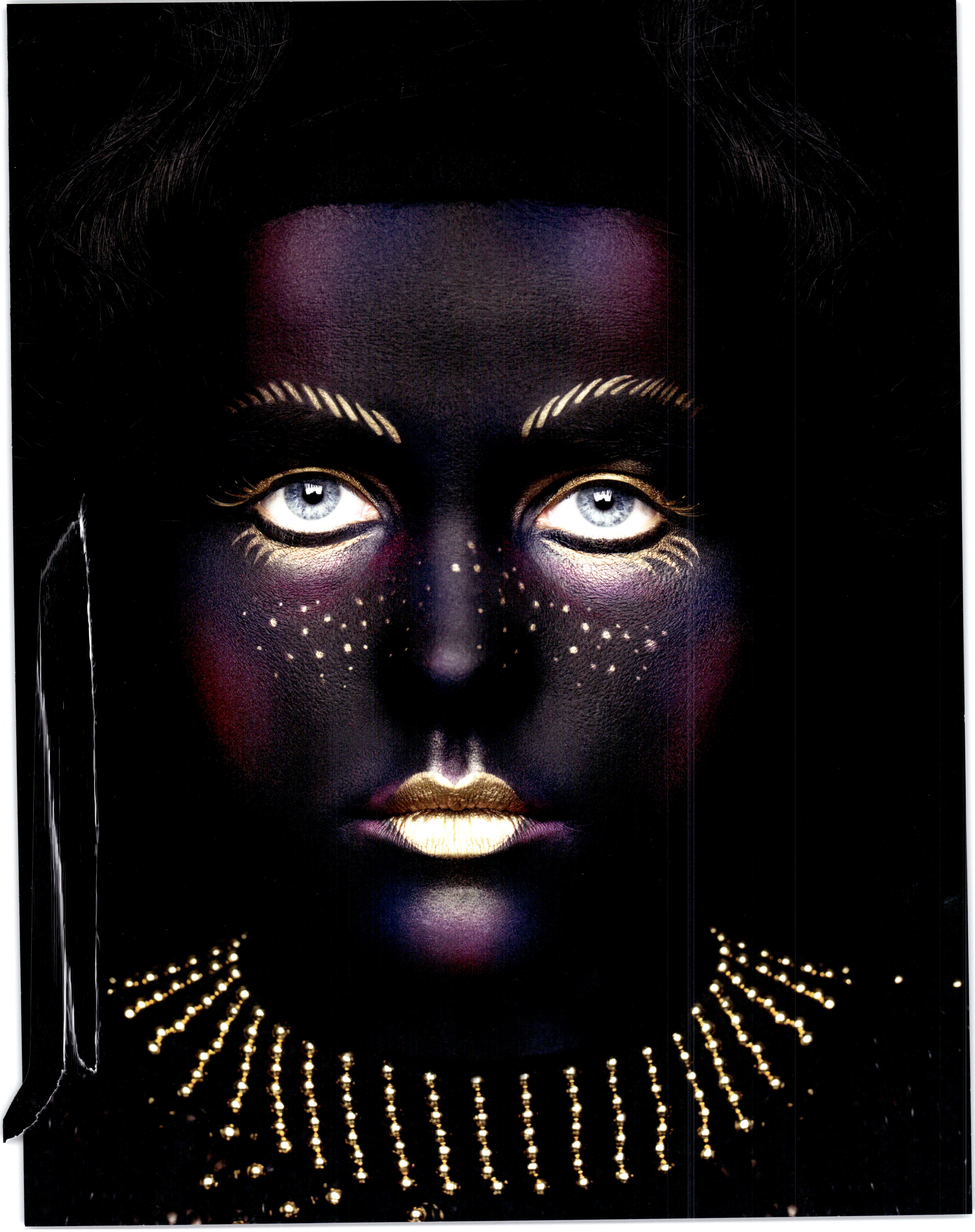

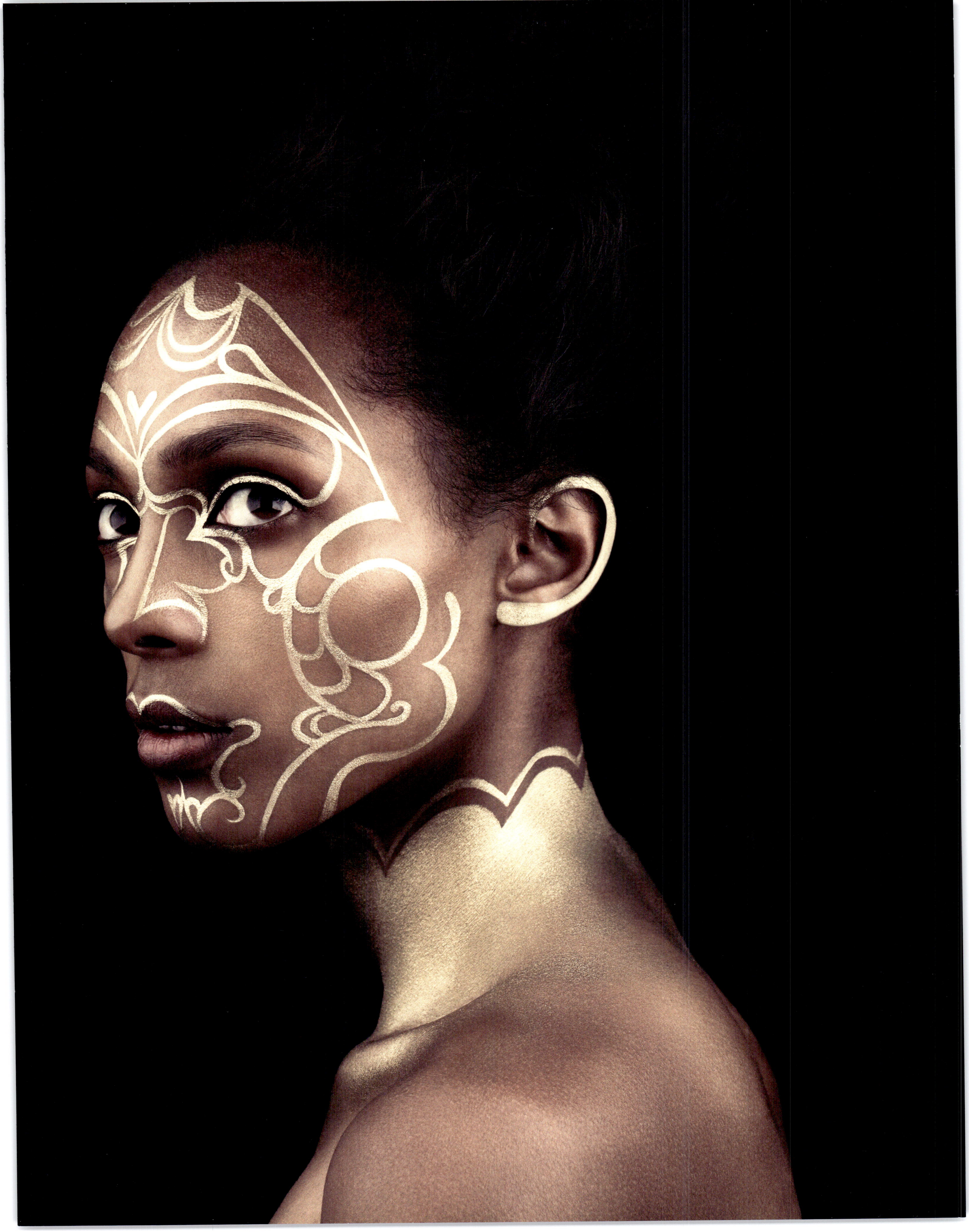

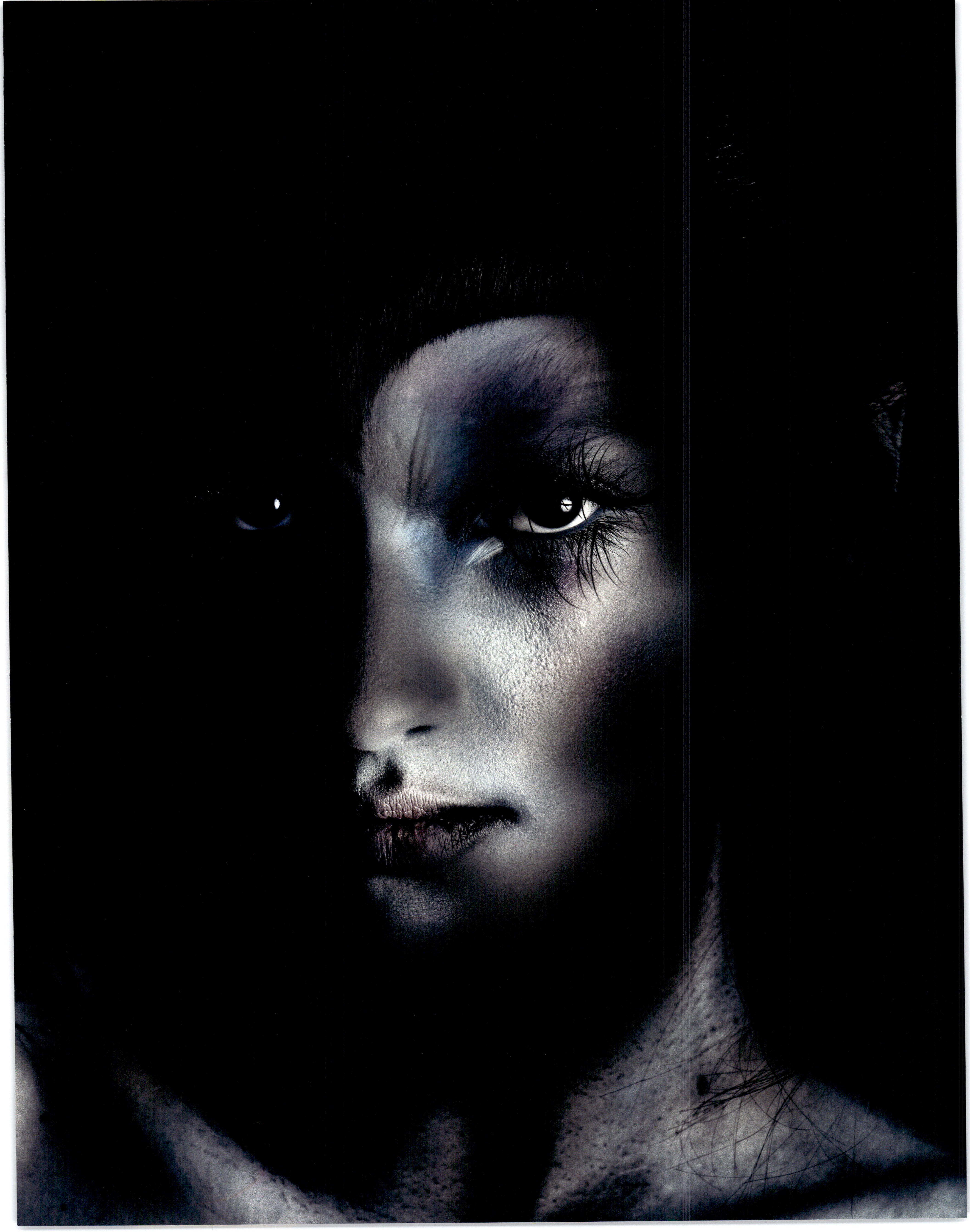

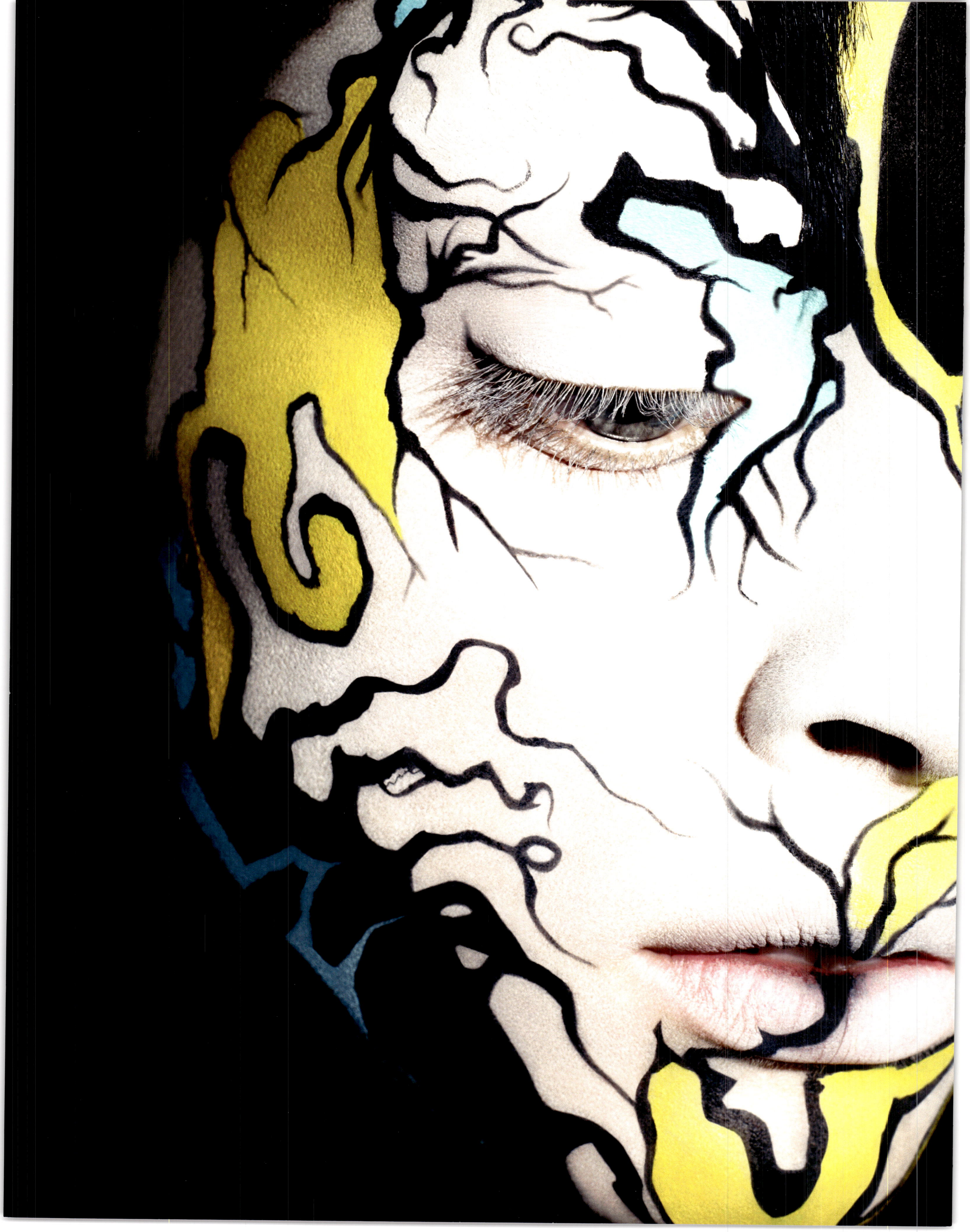

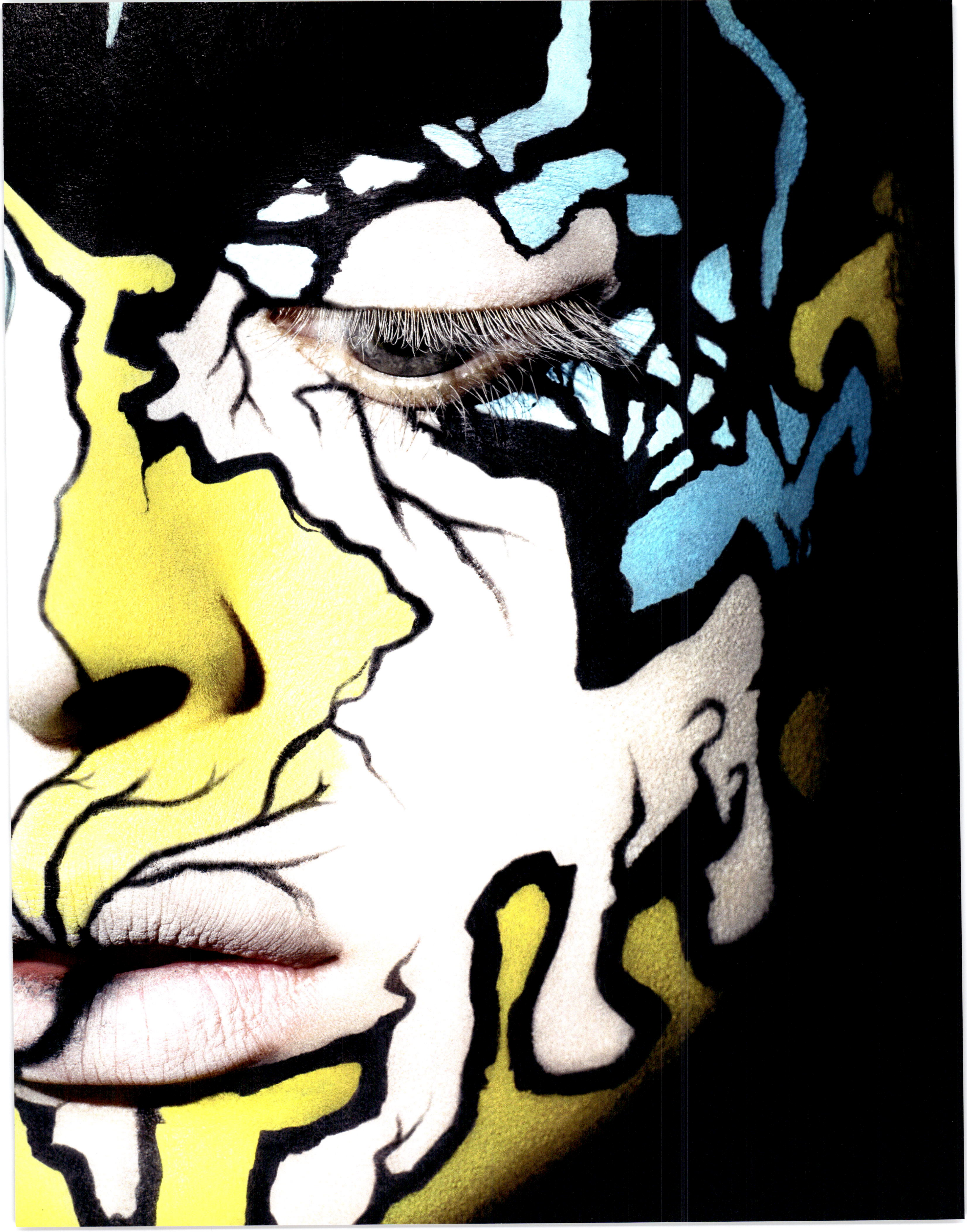

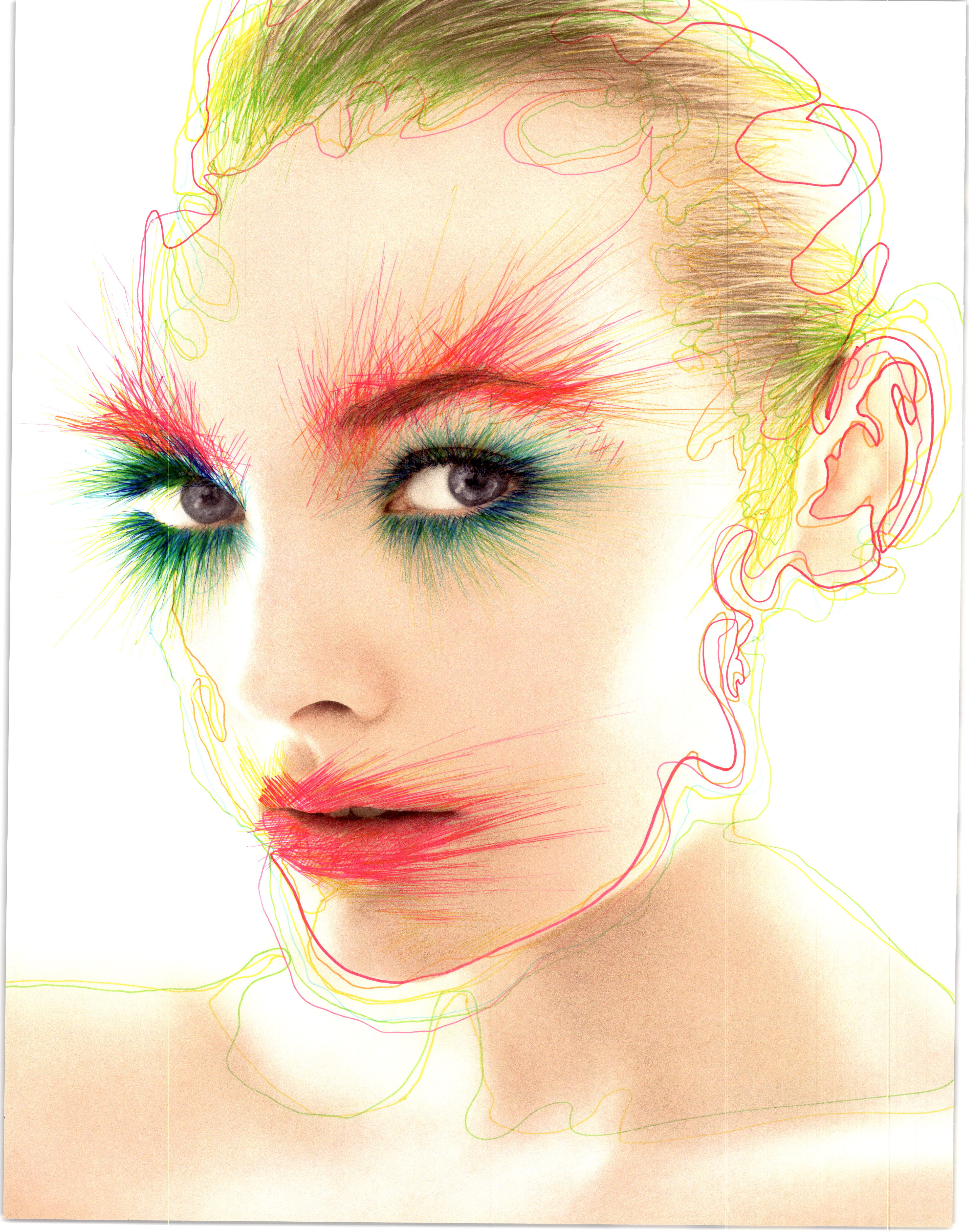

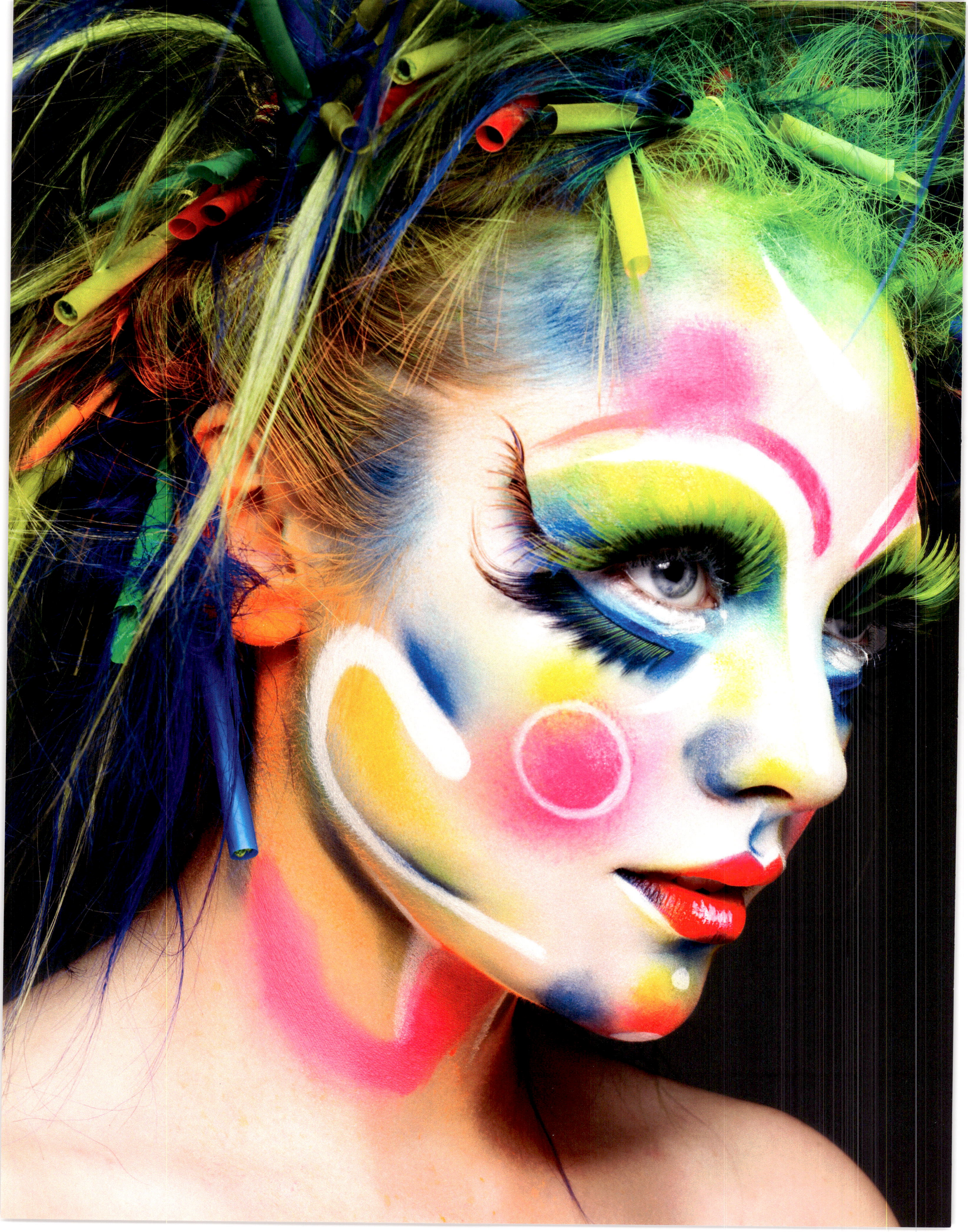

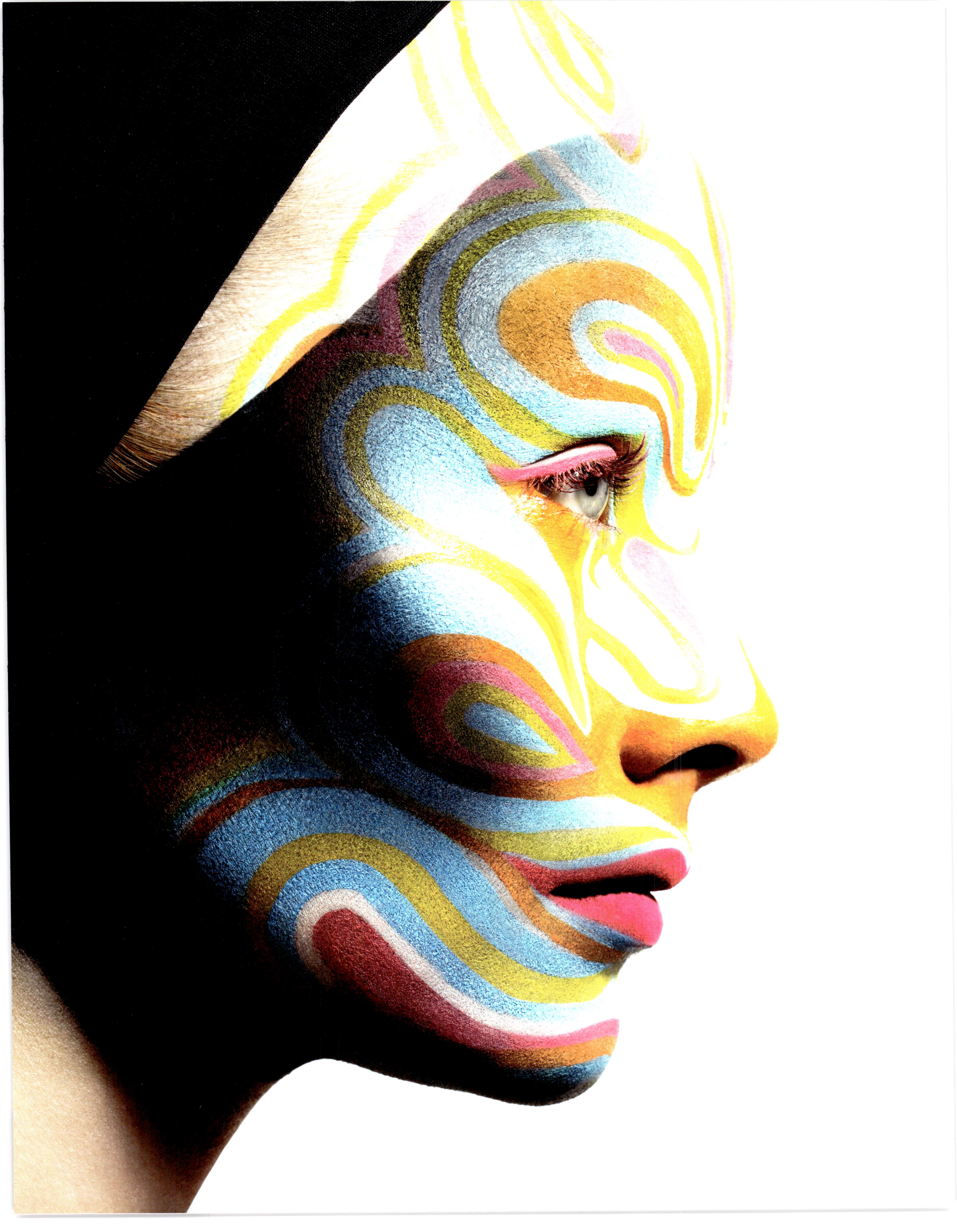

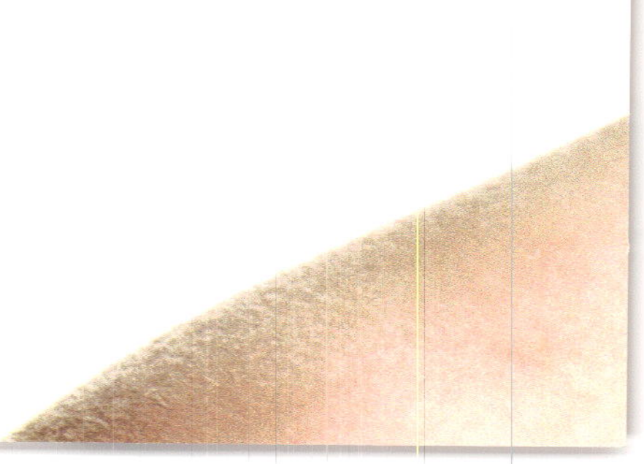

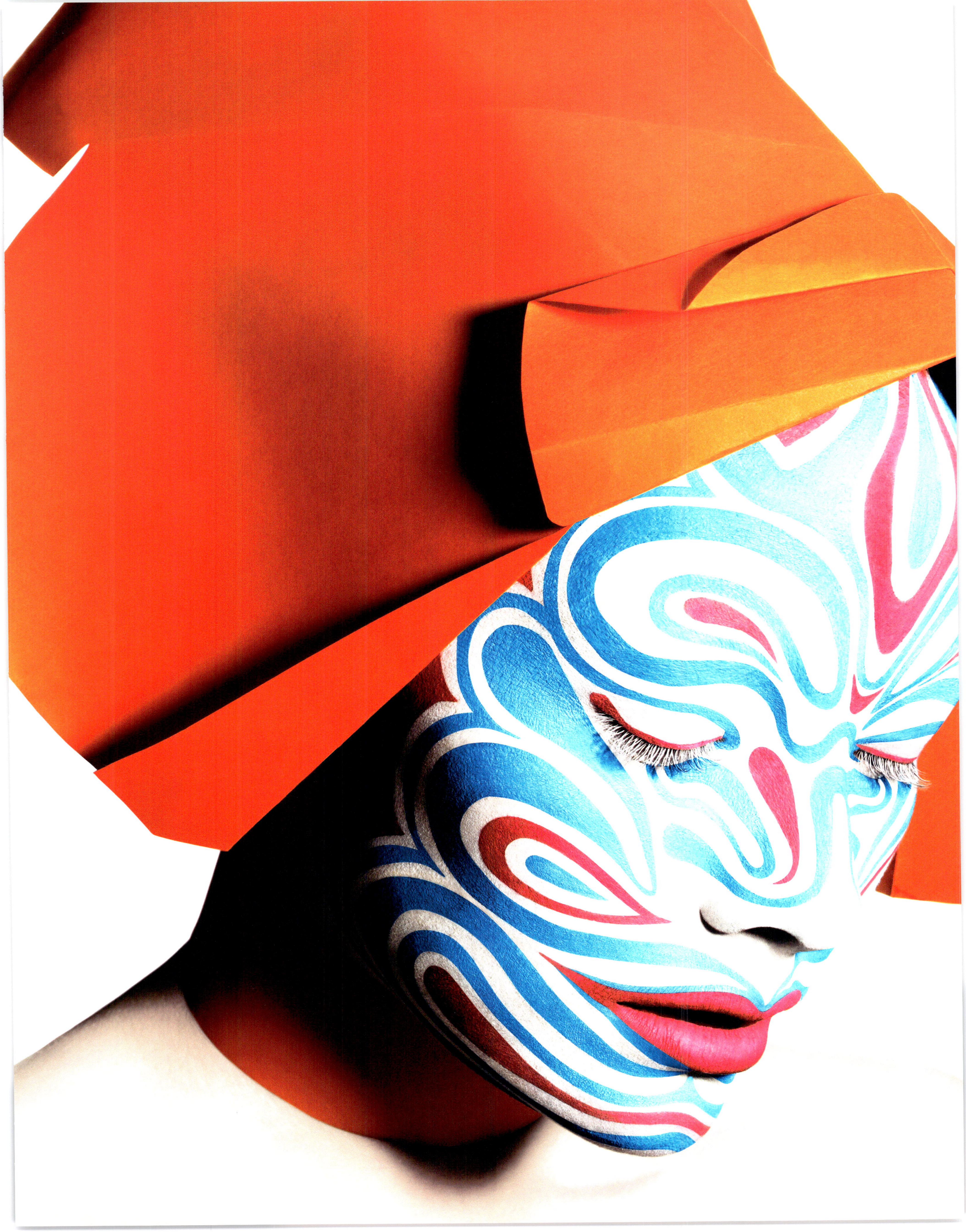

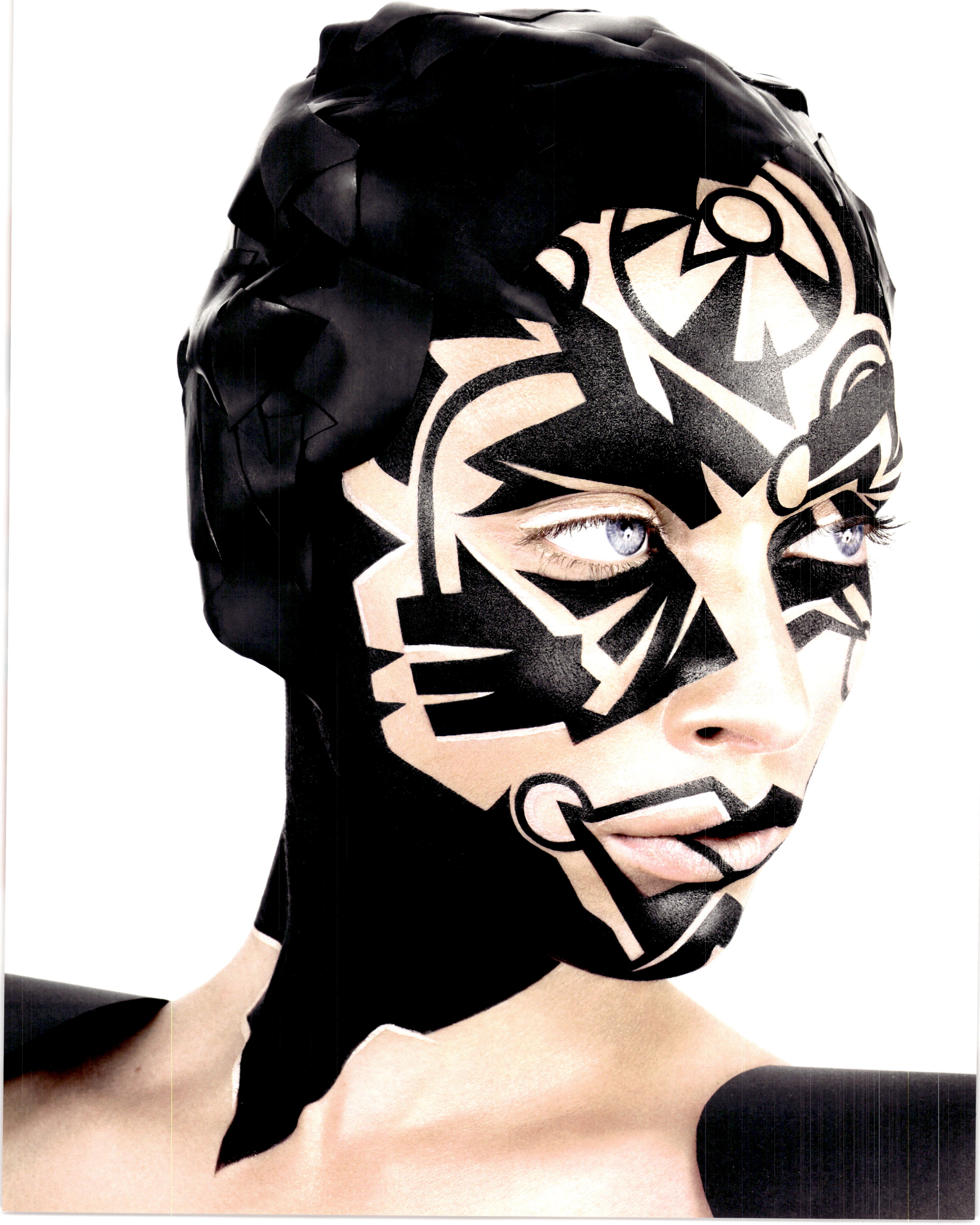

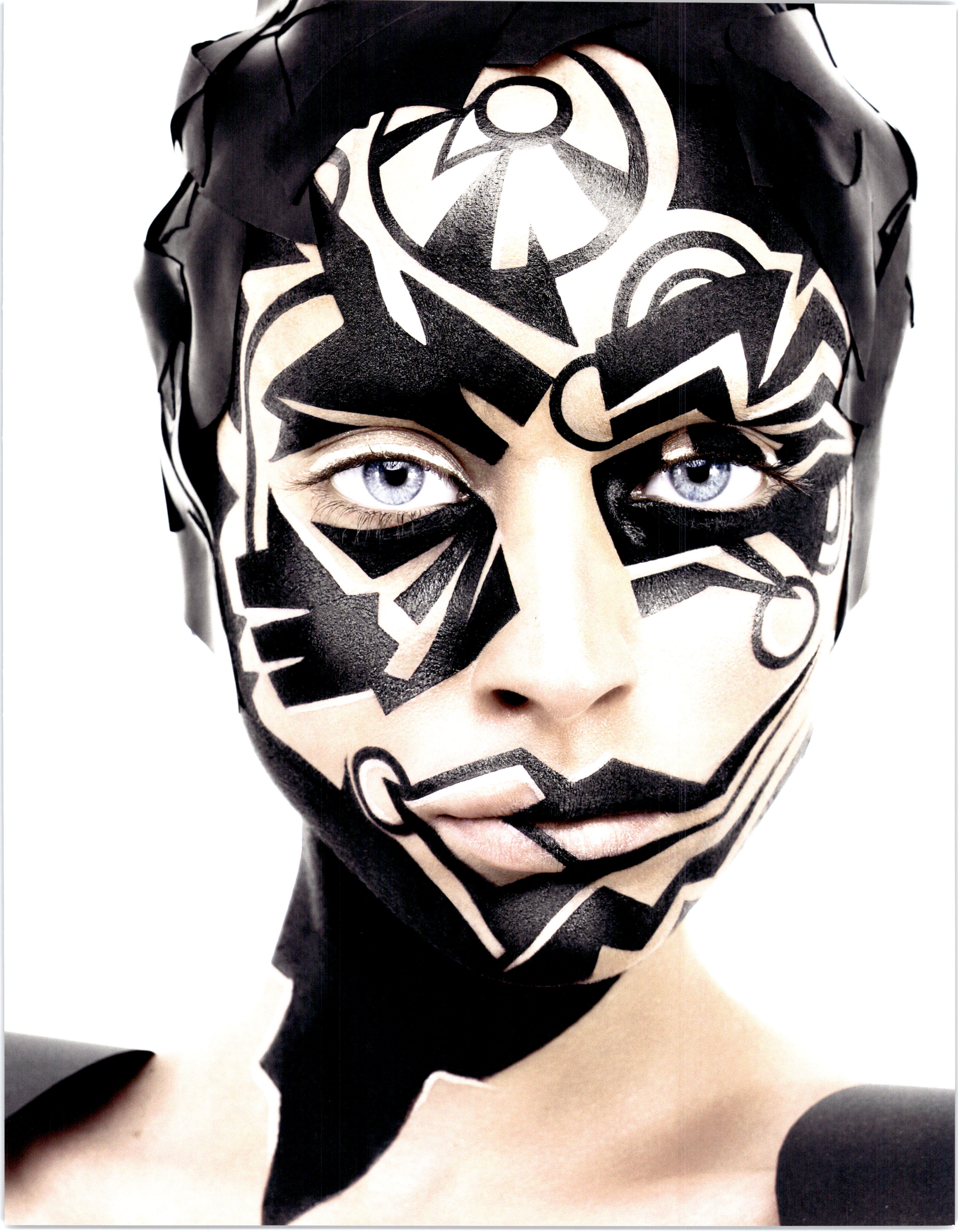

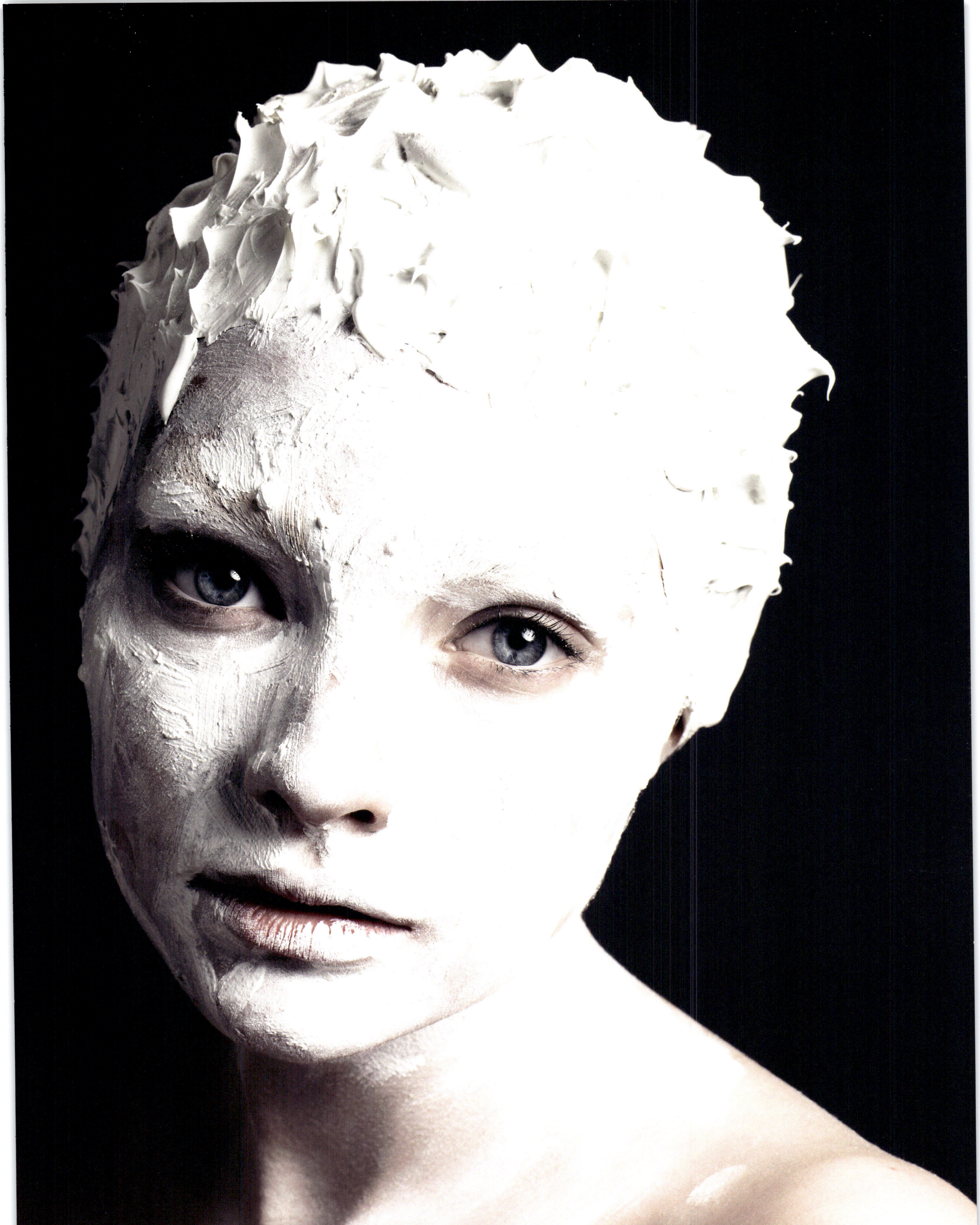

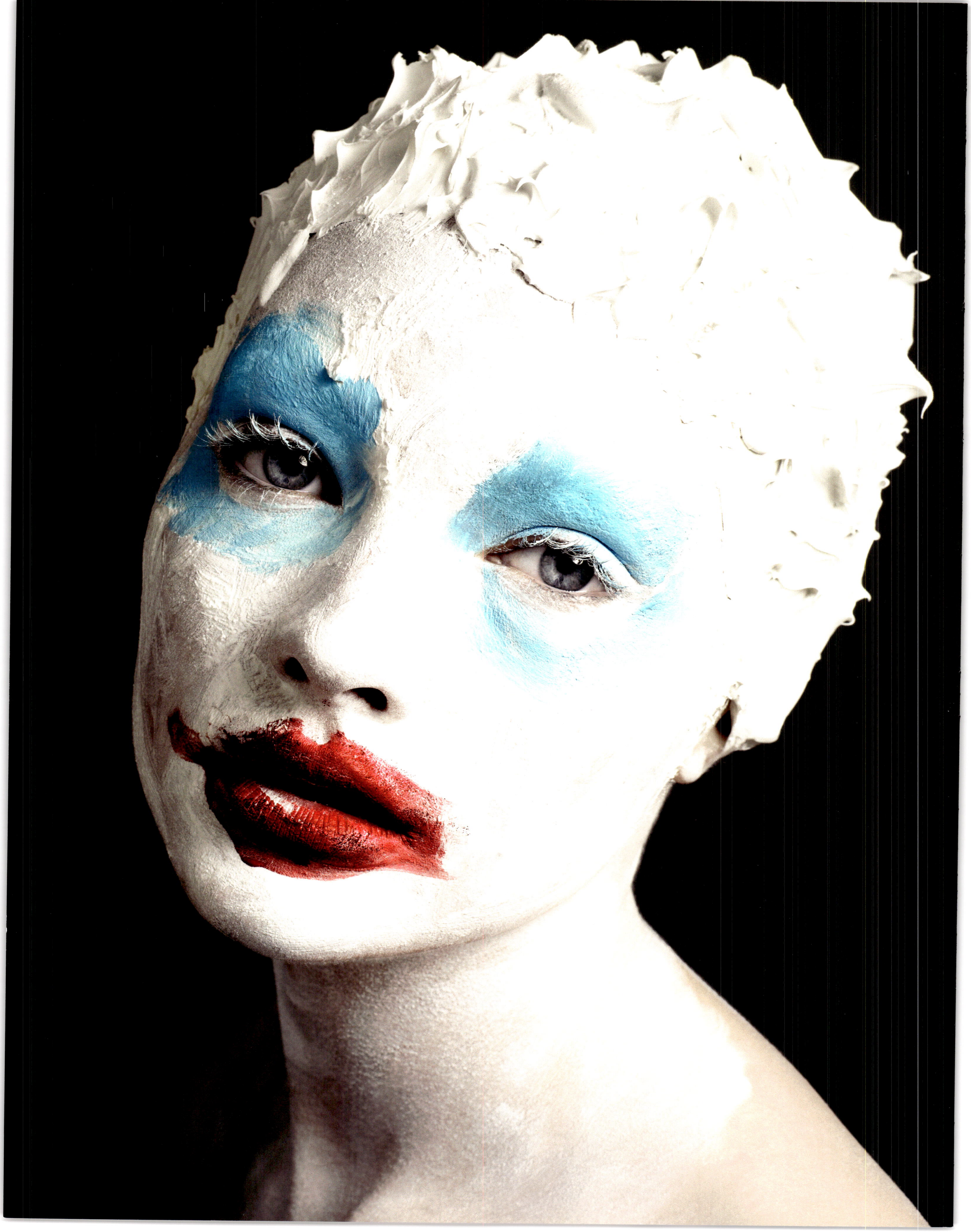

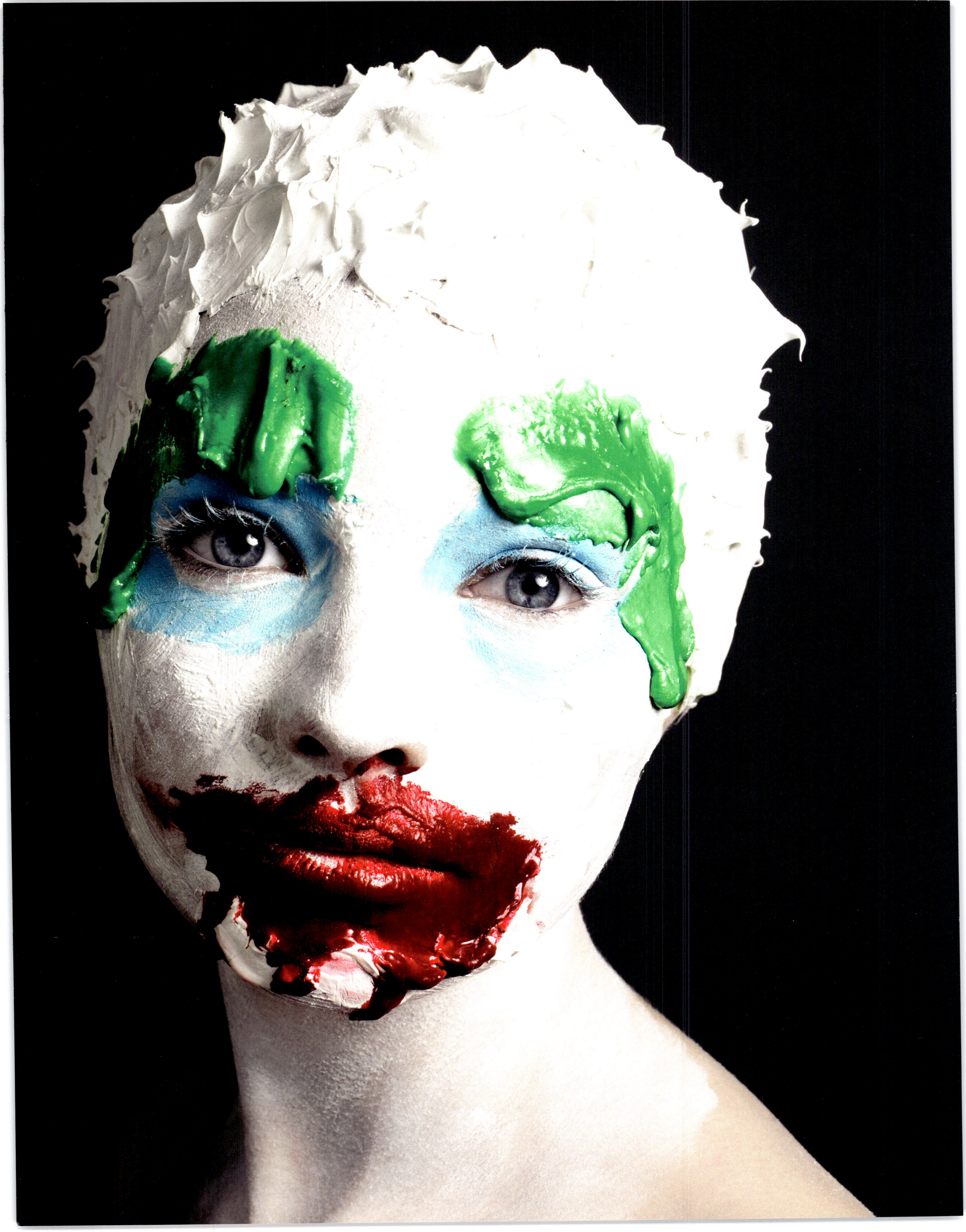

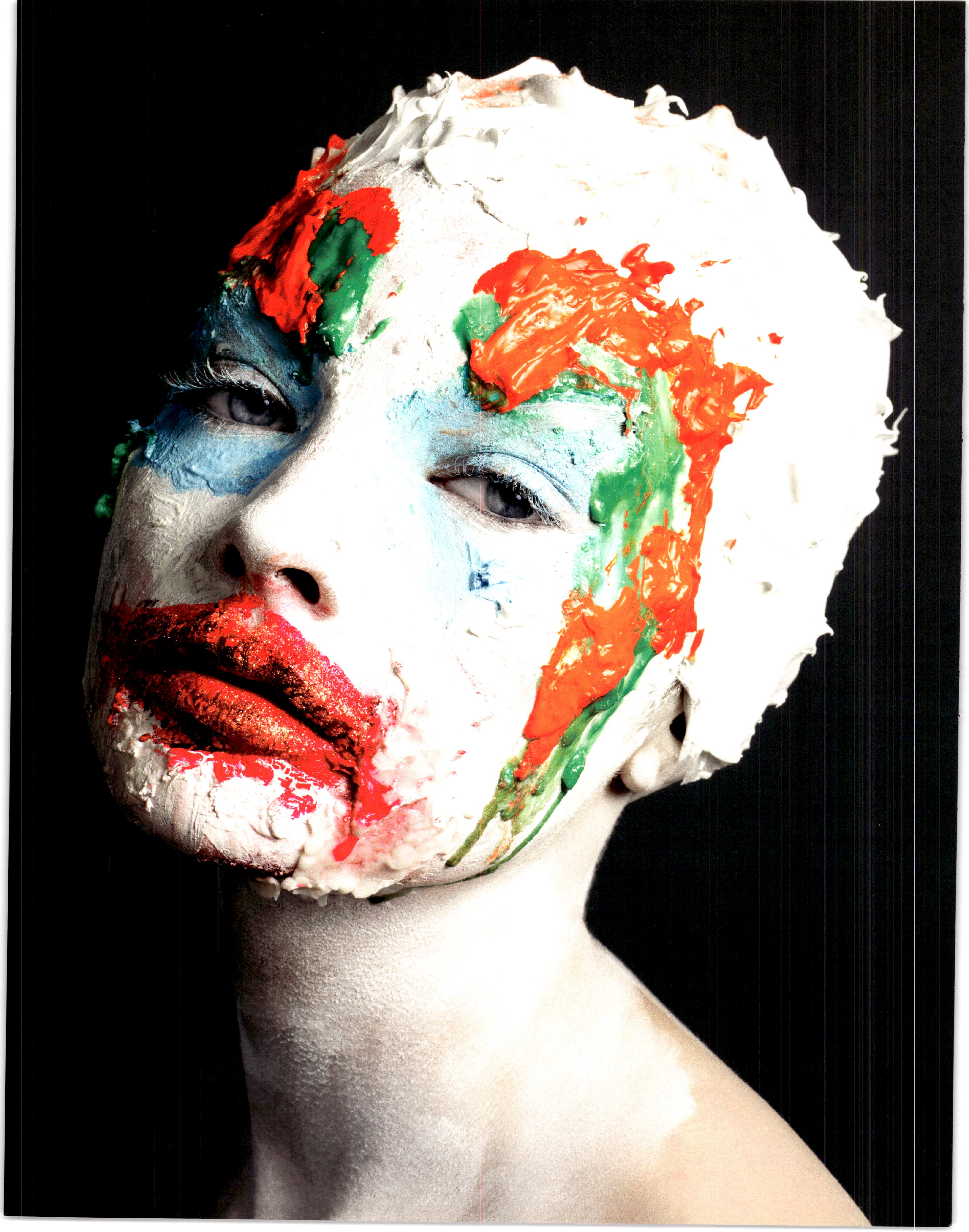

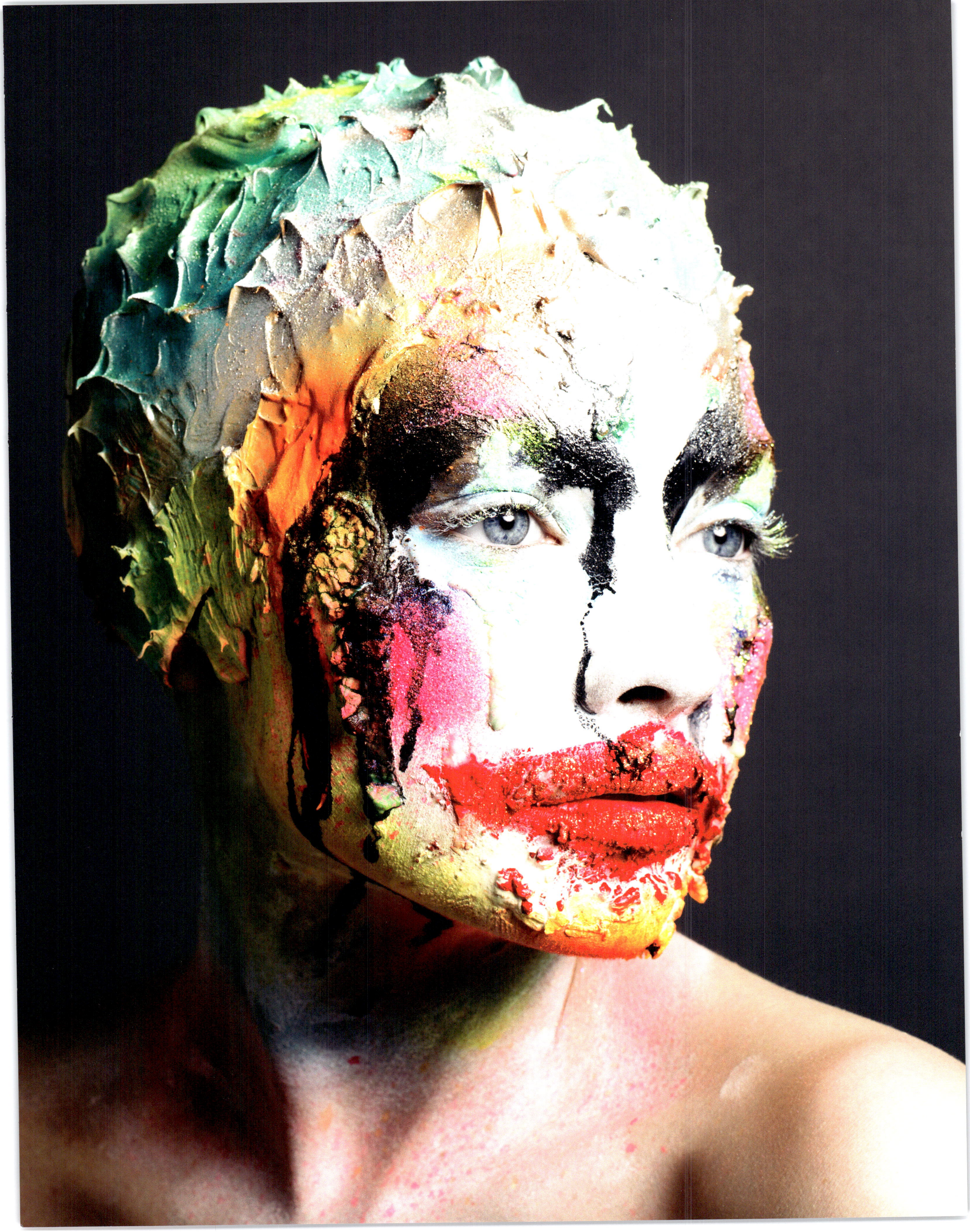

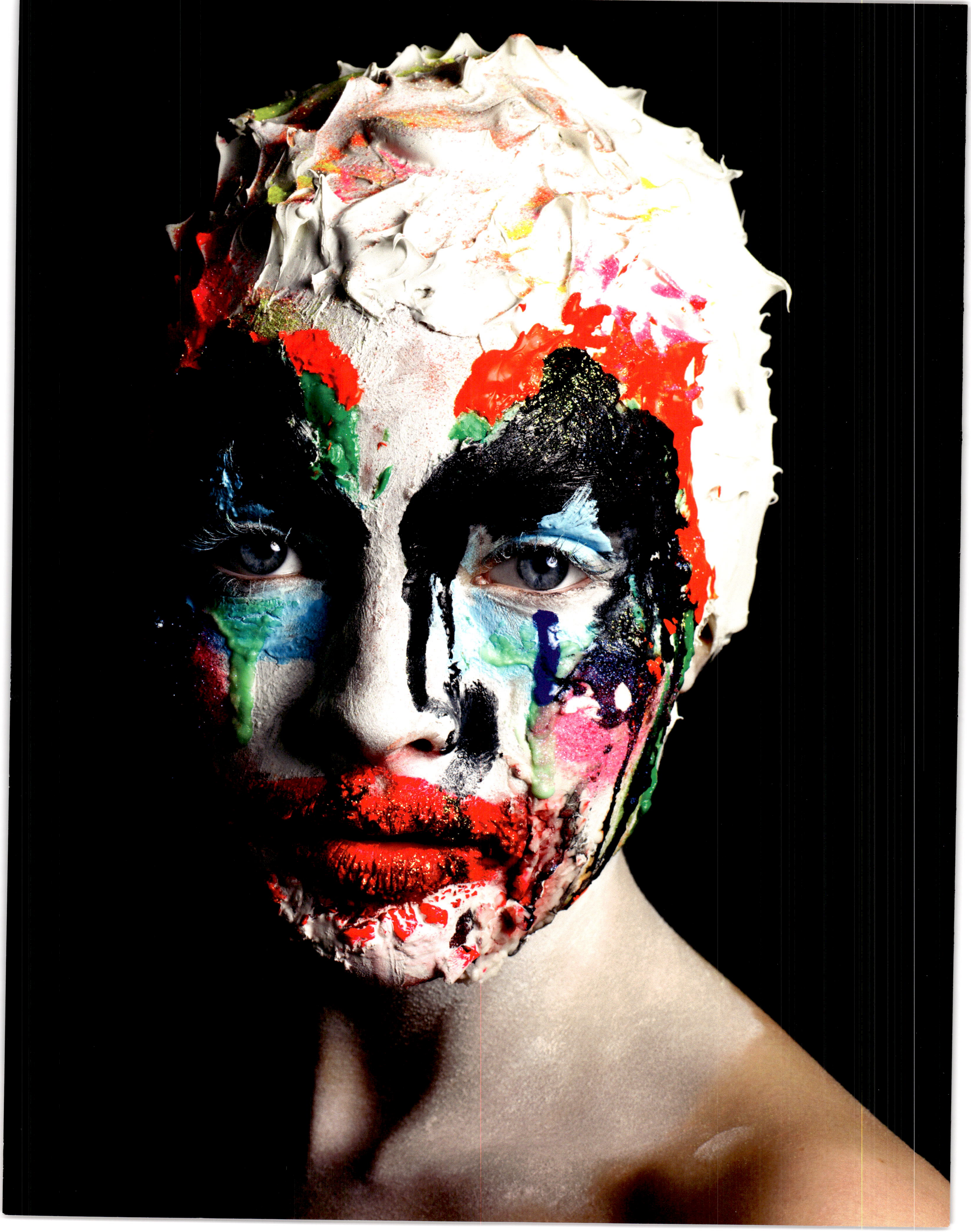

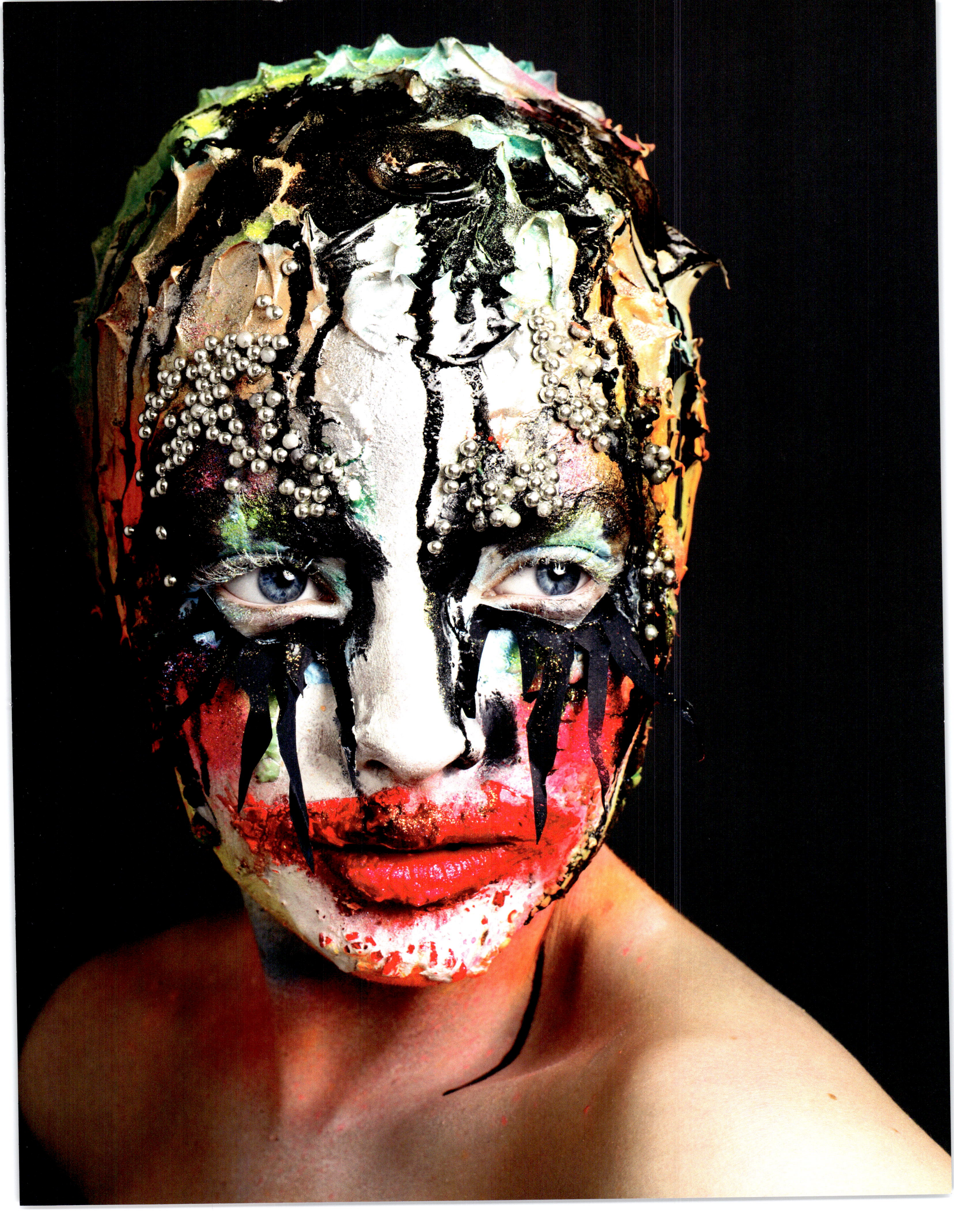

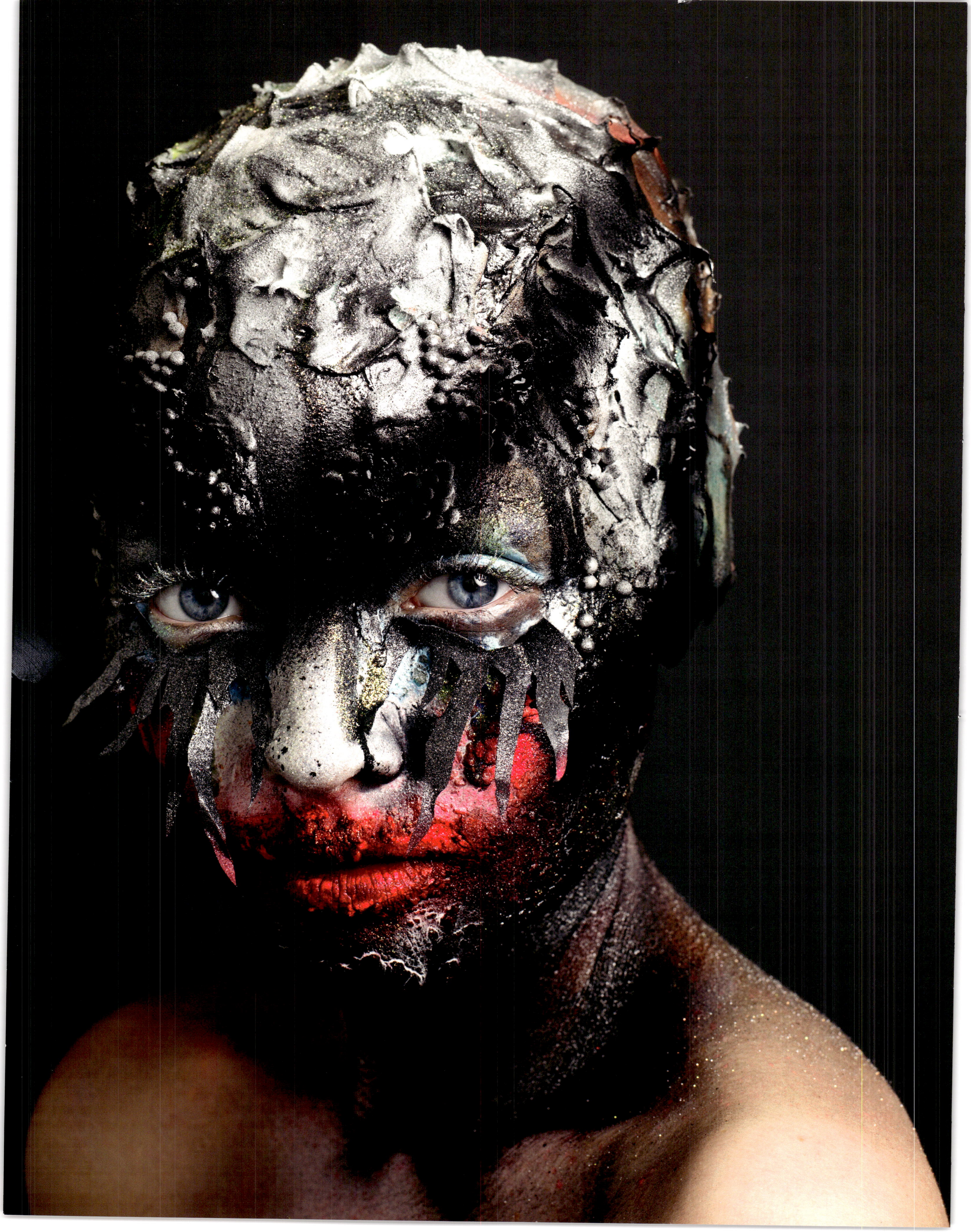

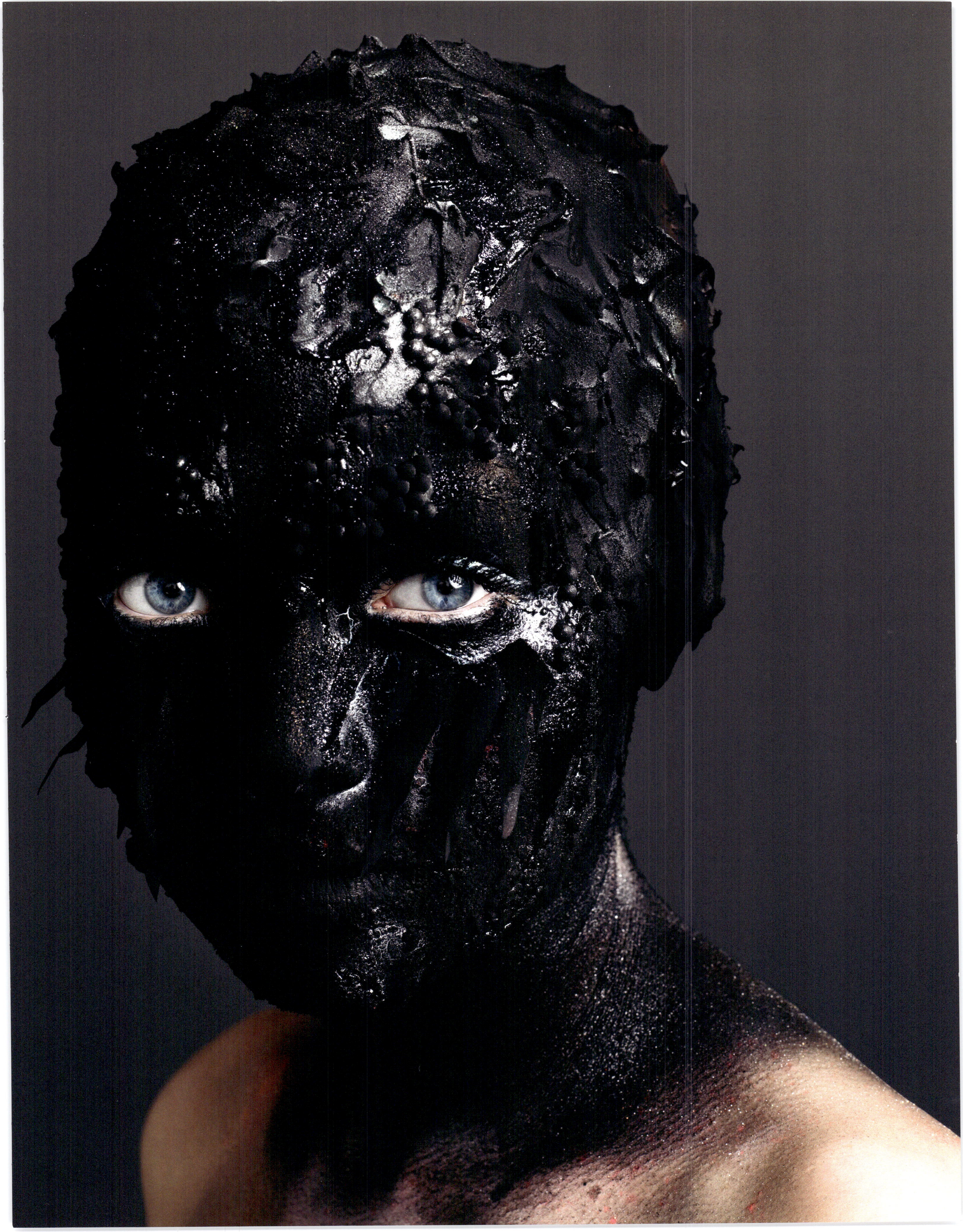

What's your first memory of make-up?

I remember thinking my mum's make-up routine was really interesting. It was such a ritual and it seemed so private and complicated. She made her eyelashes insane and huge. But, I was also influenced by her sister and my primary school teacher – all three looked really similar. My aunty Shirley, my mum and Mrs Broomfield were so Joan Collins – short hair at the back, curly on top, loads of gold, loads of diamonds, dark tanned, big eyes, lips, blusher and really long red nails. When I look at my hands and my jewellery, it comes from that kind of place. When I don't know what to wear I think, 'What would Joan Collins wear?' I've always got a really specific look, whether it's goth or something else, there's always something in the back of my head which is stuck in the Dynasty era.

When did you first think about make-up as a career?

Probably in my final year studying fine art, it was becoming very much more performance led and about bodies. Not necessarily performance, but using my body in pictures, in sculptures, in film and/or warping images. The whole time I kept scrapbooks of black eyes, green eyes, smokey eyes, red lips… all cut out from magazines. At the same time, a lot of my artwork was about breaking down advertising images. How the accents of desire are portrayed in advertising, whether it's the bead of sweat ricocheting in slow motion or the Timotei rush of hair and water or the moisturiser in a perfect shape. I started making things with skin, using body colours, advertising images and incorporating make-up into that. Then a friend of mine asked if I'd ever thought about designing for dancers or for catwalk and put me in touch with a group of dancers. They were predominantly black girls with really amazing muscular shapes. I was painting and drawing, taking out their hair and feet with cut up stockings and painting up their legs – it was a bit Gareth Pugh actually. So I started to think about make-up. I'd also started working at the Body Shop on Saturdays and then after college I worked at Shu Uemura to earn some money to pay for my artwork. I didn't know about make-up but I wore so much of it – I was such a full on goth at the time.

So what age were you when you become a goth?

When I was 16 or 17. Before that I was a psychobilly. I was always into something, not dogmatically following ideas but I was always interested in the look. I was never particularly into the music. I was really into the Cure and Bauhaus but I was also into rave music. But for me it was always about the look. I always think someone's watching even if it's just when I'm in Sainsbury's, I'm always putting on some kind of a mini show – even an audience of one is an audience. I like to get a reaction whether it's good or bad, indifference is hell – I don't understand it. I love vampires and I love the idea of Romanticism. I was very into poetry from the age of about 11 but I also loved all the Anne Rice books. Being a goth was a great way of hiding, of being really full on but being able to hide behind inapproachability – very visual but unattainable. I still like that now.

Why do you think that is?

It's protection definitely. It's a good filtration system. You can present an image at all times of intimidation or authority. It gets more extreme as you get older – I always think of Joan Crawford, the older she got the more powerful she got. The more extreme the shoulders become and the lips and the eyes, and I definitely see that in myself. Living in Grimsby, I absolutely didn't want to look like those people. I thought it was a badge of intellect in a way. 'I don't want to look like you, I don't want to live here, I want to move.' It's quite funny because when you're a goth you think you're the only goth in the world but it's actually such a uniform.

When did you leave Grimsby?

My Dad brought me down to London from an early age and we used to stay at my uncle's flat in Camden. It was always in my head, 'I've got to move to London.' When I got into Chelsea I saw it as a real coup and that sense of achievement really paved the way for things to happen. I was there for four years and was the only person doing installation in painting which at that time was quite a new thing. Frieze had only really started hitting the headlines when I was at college in the early 90s. All those guys from Goldsmiths – Damien Hirst and Rachel Whiteread were only just being touted and then in my last year Dinos Chapman came in as a visiting tutor.

How did your career evolve?

I went to work at Thomas Neals in Covent Garden. Windles, the hairdressers was also there, where most of Eugene Souleiman's assistants were. Rudy Lewis used to come over and ask if any of us were testing. I didn't know what testing was. I was surrounded by a whole new world that was connected to editorial. We started doing a few tests and then eventually I started doing stuff with Nicola Formichetti. Then i-D ran this beauty story and that got me an agent. It all started because I'd been working in Covent Garden and met this amazing group of people.

How would you describe your style?

An emotional response to an impetus – whether it's the direction of the photographer, the stylist, the designer or, like doing the book, something that comes from me that day. I like to react to the person in front of me. You have to let the face lead you. I find my self very affected and it feels like painting. Doing installation art can sometimes feel like you're a brickie or a set designer, you get bogged down in the mechanics but actually painting is so much more responsive and in the moment.

Q&A with Alex Box

How did you first come to work with Rankin?

I met him quite a long time ago. I think he saw something in me that he saw in himself when he started. I've seen the stuff he did when he was at college which was much more experimental. I think that Rankin loves to surround himself with people who remind him where he came from and what's he's about essentially. He's also got a very crude sense of humour like I do, he didn't scare me – if anything I think we give each other as good as we get. He's high octane, he needs to be charged by something – people need to give him something to chew on.

Do you think you share the same views on beauty and how do you think that relationship works?

I think we are very different. Yet, we are both naturally very curious. He's quite anthropological and I think he sees me as one of his subjects. All photographers view their world from behind their lens – it's like a petri-dish situation. It's an experiment

How did the idea for the book begin?

Dazed did a beauty issue and Rankin asked me to do a beauty image with him. It was a really black, linear image – it's in this book. He was really excited by it and wanted to do more anthropological work like this. I'd got one of his beauty books and because it was close-ups of skin, I would draw on make-up ideas. When Rankin saw it he couldn't believe I'd drawn all over his book, but then the idea grew and he said we should do a book together. That's why in the book it's a mixture of real time images and drawings.

Did you and Rankin discuss ideas prior to each shoot for the book?

No, I never knew what I was going to do and neither did Rankin. I'd make it up on the spot, I'd draw ideas when they came to me and collect images, but it would always change when the model sat in front of me. It's interesting for me to see Rankin's eye on what I do, he has a quite childlike curiosity. We fashion types can tend to get wrapped up in things being very intense. Somebody has a clown face but it's all very serious, but if you see it from the outside it's hilarious, ridiculous, verging on complete nonsense. But Rankin can take a picture very quickly and be very subjective. It needed to be like that – like a living diary.

How did you approach the images where you applied the make-up as drawings or collage afterwards?

Rankin would take pictures of different head positions and I would choose a few, get them printed up and draw on them. Almost like homework, crafternoon – it's such a lovely thing to do. It was a real pleasure to be able to start drawing again but with a purpose. What used to frustrate me about artwork is that it needs to have a purpose and it needs to be seen – I'm an only child and I'm quite showy. I now had the impetus of the 2D face, and could play with exploding the boundries and dimentions. It's freed an art block I've had since I left college and even though I'm a make-up artist now, my work has become more cohesive. I've come full circle – it's interesting that I started off as an artist doing make-up on my art and now I'm a make-up artist doing art on my make-up.

Do you have a favourite image in the book?

It's the darkest one, black veins, black tar in the hair, everything's black, black. There's a lot of veins in the book – I think it's very me. Other than the one I did with Rankin for Dazed, which I absolutely adore and stand by as absolutely the Box. The image is cinematic; it looks really otherworldly, poetic and quite sad. The model looks very involved in the moment – it's quite sombre and very quiet.... and I keep saying very quiet, people will say it's so full on. But I never want the model to look anything but quite distant, quiet and a little bit tragic. There is humour in there but it's a dark sense of humour. The girl is always empowered – I never like to see images of victims. Some images that I've seen in the fashion business show models weighed down by the hair and make-up, but with me I really strive to make sure that the person is still breathing through it and you're still connected to their spirit. If anything, they're lost in the moment. The minute you obliterate somebody's spirit, you're abusive. That's a strong word to use but there's so much misogyny in fashion, people are so quick to make models little girls, hookers, broken dolls.

So, you're experimenting a lot with textures and the actual medium of make-up?

I'm obsessed with Francis Bacon, I always think I want to Bacon it up somehow. For the last shoot we did for the book I was thinking about three dimensional make-up and things sliding into one another. I used set icing sugar on the face and textured paint on the head. I still don't think I've even skimmed the surface of what I can do – 'must try harder' is always the doctrin in my head. I purposefully didn't do any masks or anything appliqué in this book because that's another book altogether. The linear line is very apparent and there's not much stuck on. It's about medium – powder, paint, drawing and ink.

Do you have a favourite feature of the face?

Eyebrows – I'm always taking them out and then putting them back in how I see fit. My ideal person would be someone without eyelashes and eyebrows because you get to see exactly who it is under there. When you take away eyebrows you get that beautiful bone and it takes away a lot of expression and you're left with just the architecture of the face. So many of the images that I love are from the 18th century – where there is no facial hair and it's just dark sockets. The National Portrait Gallery has inspired my make-up for years.

How would you describe the book?

An emotive make-up diary. There are points when I can see I was sad and others when I was happy. It's a diary and to me it's very revealing.

Alex Box Epilogue

You will probably have noticed that, until now, there haven't been many words in this book. That was intentional, as was the decision to forego any kind of introduction. Instead, there's this, an epilogue, the idea being that the viewer gets to look at the images then decide for themselves what they think without the skew of someone else's opinion or the interpretive "lens" of a commentator's analysis trained on each picture to guide their reaction. Alex figured that if you were sophisticated enough to pick up this book in the first place, then the last thing you would expect is to have your hand held as you turn each page, reading explanatory notes and insights regarding the artist's innermost thoughts and the tools and techniques used in the creative process. Knowledge can spoil things, can't it? So don't expect any step-by-step explanations: makeover manual this ain't.

What is important is the subjective appreciation of the images you have seen. We all process visual information in the same physiological way: photons hit the lens in the eye and are focused on photo-receptive cells in the retina. Incoming light is converted into electrical impulses that are sent via the optic fibre to the visual cortex of the brain. The assessment of any image then goes through a subconscious filtering process, with each layer designed to further distill and refine the visual information so that ultimately sense is made of what is seen. This entirely subjective reaction happens almost immediately but is referenced in a whole realm of past experiences, prior knowledge, previous convictions… so what my brain perceives is highly likely to be different to what your brain perceives; a further reason why there is no explanation of the images as that would guide and control your perception.

Indeed, it's impossible to be impartial when considering the images in this book. They are moving, emotional, challenging, intriguing. Take page 84-85, for example, what I call the Russian Constructivist Mickey Mouse (but I'm sure you'll have your own label for it). It appeals on so many levels – the simple, classic choice of colours – black, white, red; the clever circles that make curls of hair (or are they ears?); the neat ink-outlined collar that hints at fashion; the blanking out of the model's features so that the face becomes flat, one-dimensional, so that it can be re-configured to the artist's specifications; the sardonic smile created by the sneer of unevenly drawn lips and the tug of those red tramlines that disappear off into the distant hairline like some never-ending road. And what's with the eye decoration, all mangaesque, doodleish? The more I look, the more I see: haughty Spanish Contessa, Berlin cabaret artiste, New Romantic on a night out… But stop: these are my references, not yours.

The worst criticism that's levelled in fashion these days is that your work is "referenced". Reading between the lines, this means unoriginal, that you took the idea from something or heaven forbid, "someone" else, i.e. that it isn't YOUR idea. But surely it's impossible NOT to have a reference in your head when you are creating an image. Growing up in the 80s and 90s provided, says Alex, an "open sewer" of ideas and together with these fashion "genes" other sources such as art, science, nature and magic evident in her work mean that she has evolved a visual DNA that is uniquely her own. There are nuances of hints of threads of references that become melded together harmoniously through the artistry of execution.

Then there's the process of creation and that is something else entirely. The work you have seen here has been collated over the past year and a half. It's essentially an open sketch book, a view of the workings of an artist's mind. "Sometimes it's like meditation; totally unplanned," says Alex, " as my hand simply travels across the face and the results you see here happen because the structure of the face dictates where the line or colour goes." When asked about specific images, Alex can't explain why she did something, just that it felt emotionally right at the time. Take the image on page 86-87, a dark, aggressive, worked-over portrait. "When I look at that now, I don't know what I was thinking or feeling although I recognise that the result is certainly a reflection of my state of mind."

Artists always talk about challenging our preconceived ideas, pushing boundaries, breaking rules, and I guess that taken as a whole, the kind of distortions and manipulations of each face in this book are exactly that, but – importantly – what remains and triumphs, is beauty. Her approach is purely artistic – Alex studied fine art at Chelsea and has exhibited as an installation artist exploring the relationships between the body and environment and what she terms the "distillation of desire". However, identifying and isolating what these desires are is a complicated matter. "We're overworked on advertising, always chasing a logo," says Alex in explaining the endlessly refined codes that propel us to make decisions to consume or not. "The eye seeks perfection and imperfection and if I can make someone see that both are beautiful, then I've achieved something."

Alex is fiercely motivated to champion "otherness" both in her editorial work and in her role as creative director for the make-up brand Illamasqua. Her work is the antithesis of the contemporary codification of beauty: the homogenised, plasticised, airbrushed, perfected, pimped-up "ideal" that society imposes on our collective consciousness. If that's what you like, great, but don't expect to find it here in this book

Anna-Marie Solowij

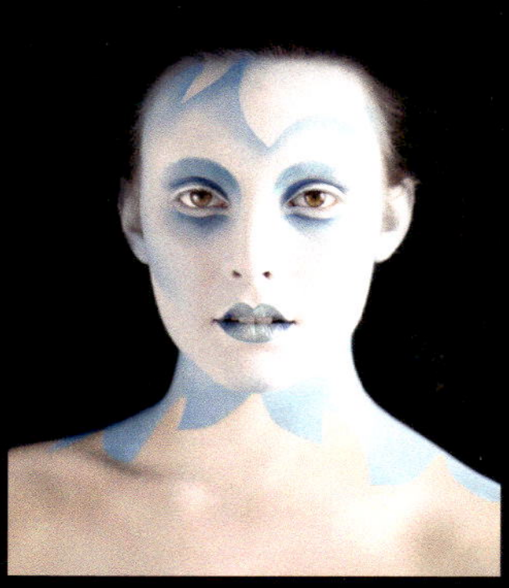

Model
Donna McPhail
Hair
James Pecis
at D&V Management

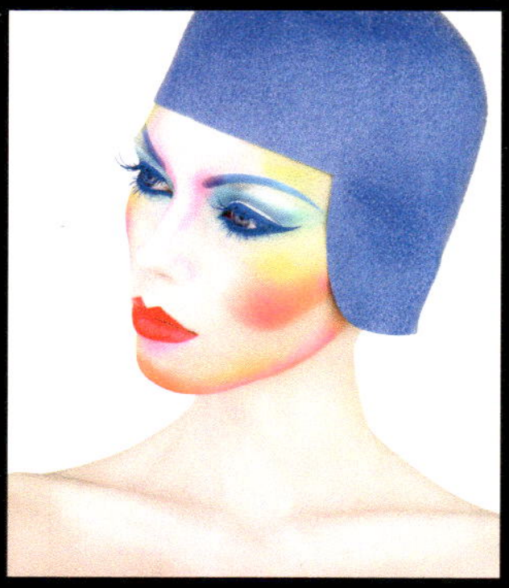

Model
Veronika
Hair
James Pecis
at D&V Management

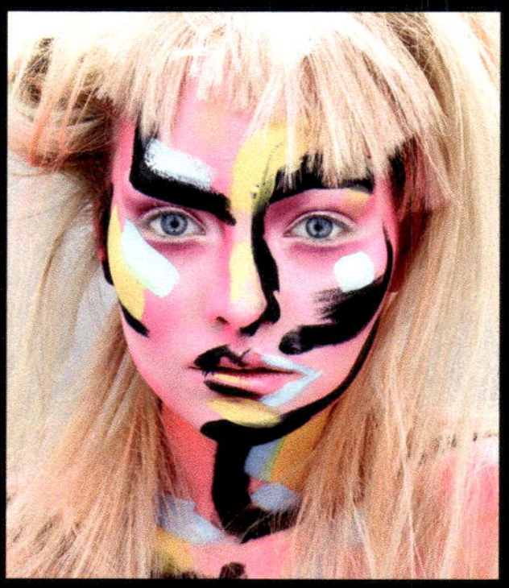

Model
Charly Wright
Hair
James Rowe
at D&V Management

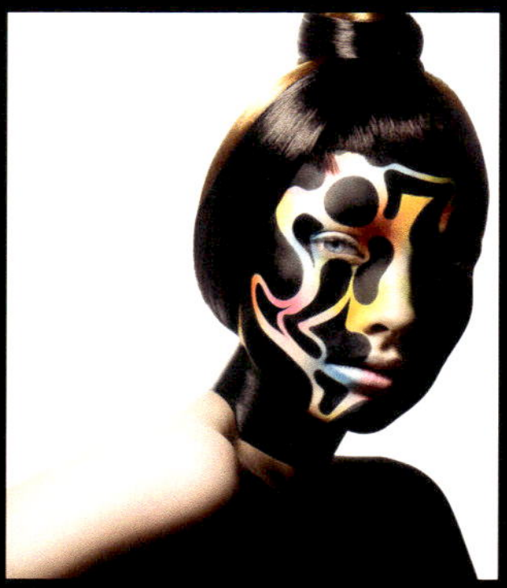

Model
Charly Wright
Hair
James Rowe
at D&V Management

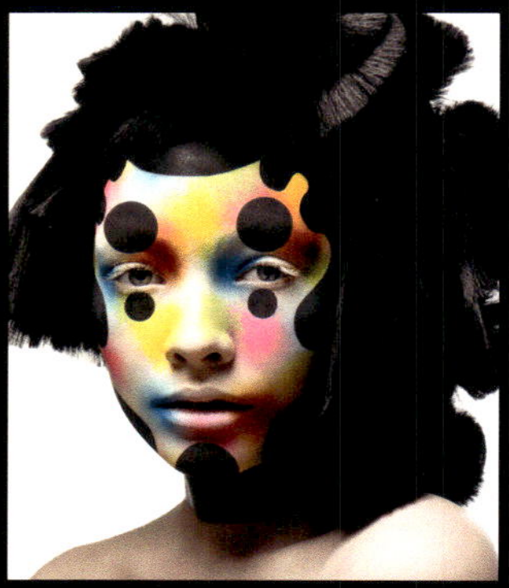

Model
Charly Wright
Hair
James Rowe
at D&V Management

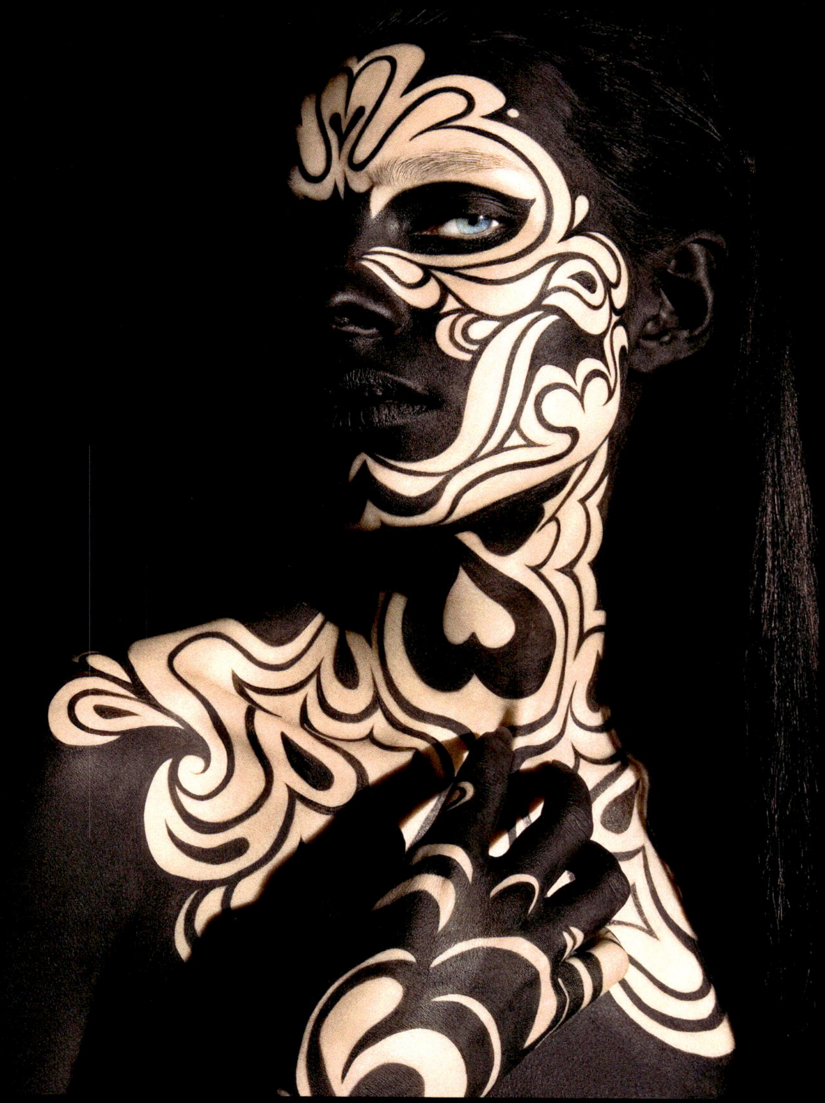

Model
Irina K
Hair
Steven Low

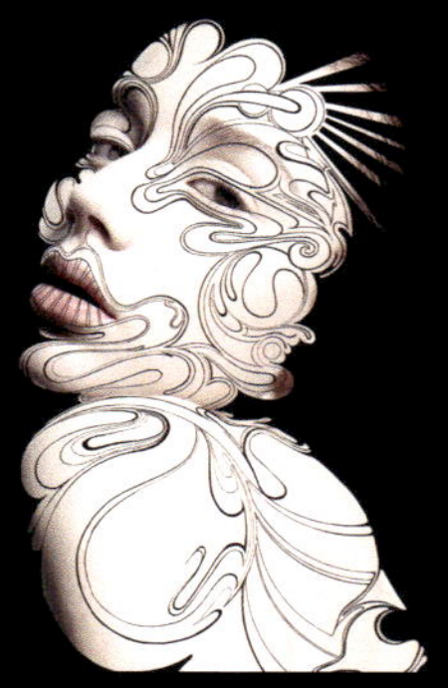

Model
Tatiana
Drawing
Alex Box

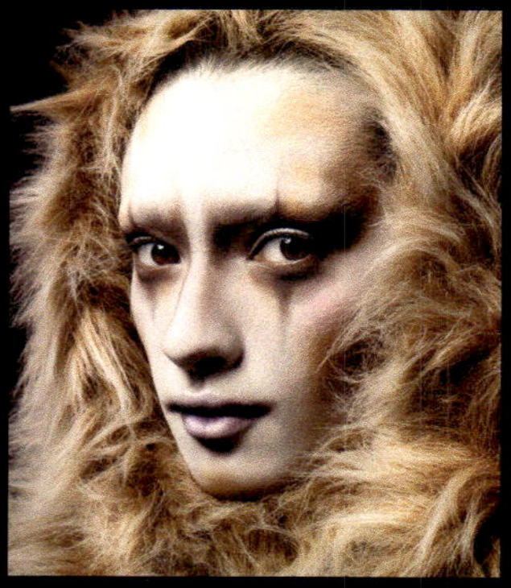

Model
Elina
Hair
Alex Box

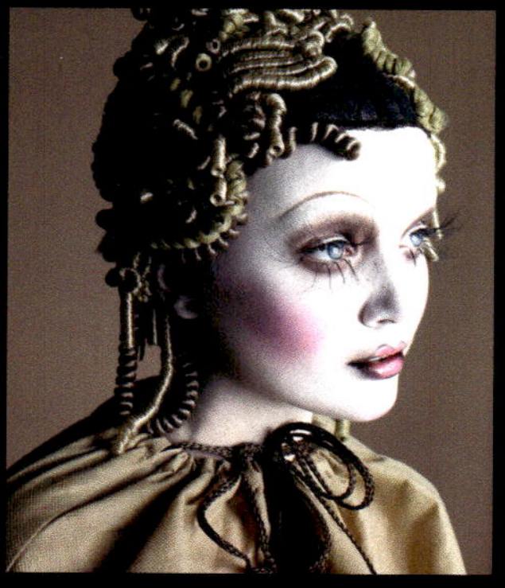

Model
Dangole
Hair
James Pecis
at D&V Management

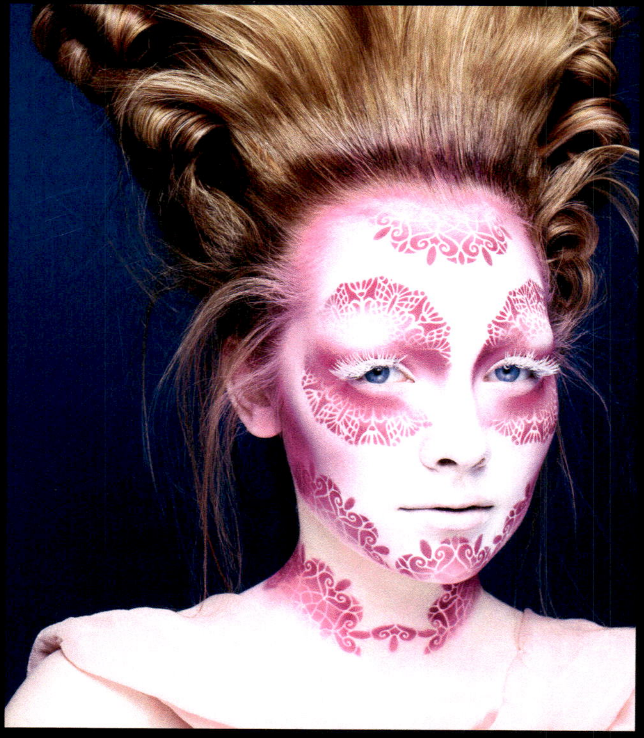

Model
Charly Wright
Hair
James Rowe
at D&V Management

Contributors

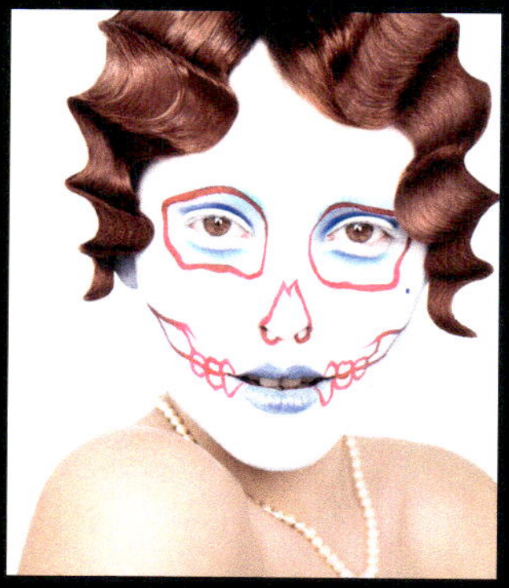

Model
Donna McPhail
Hair
James Pecis
at D&V Management

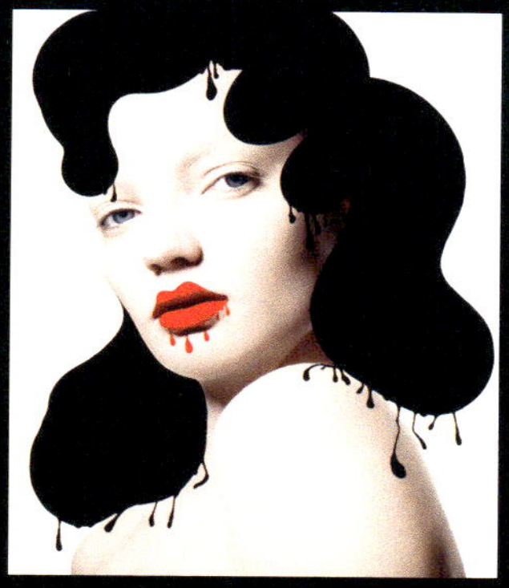

Model
Natasha
Drawing
Alex Box

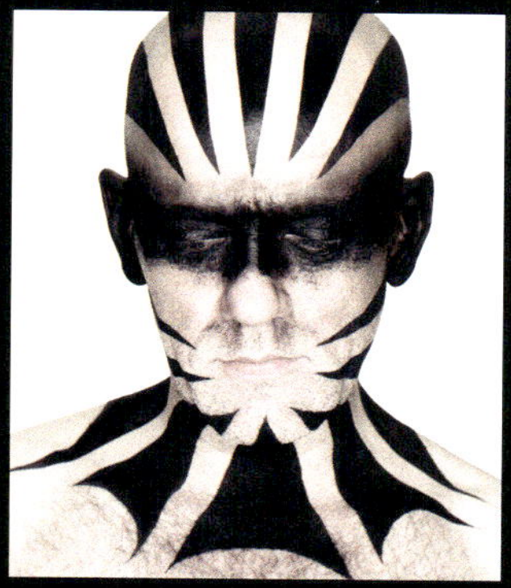

Model
Michael Stipe

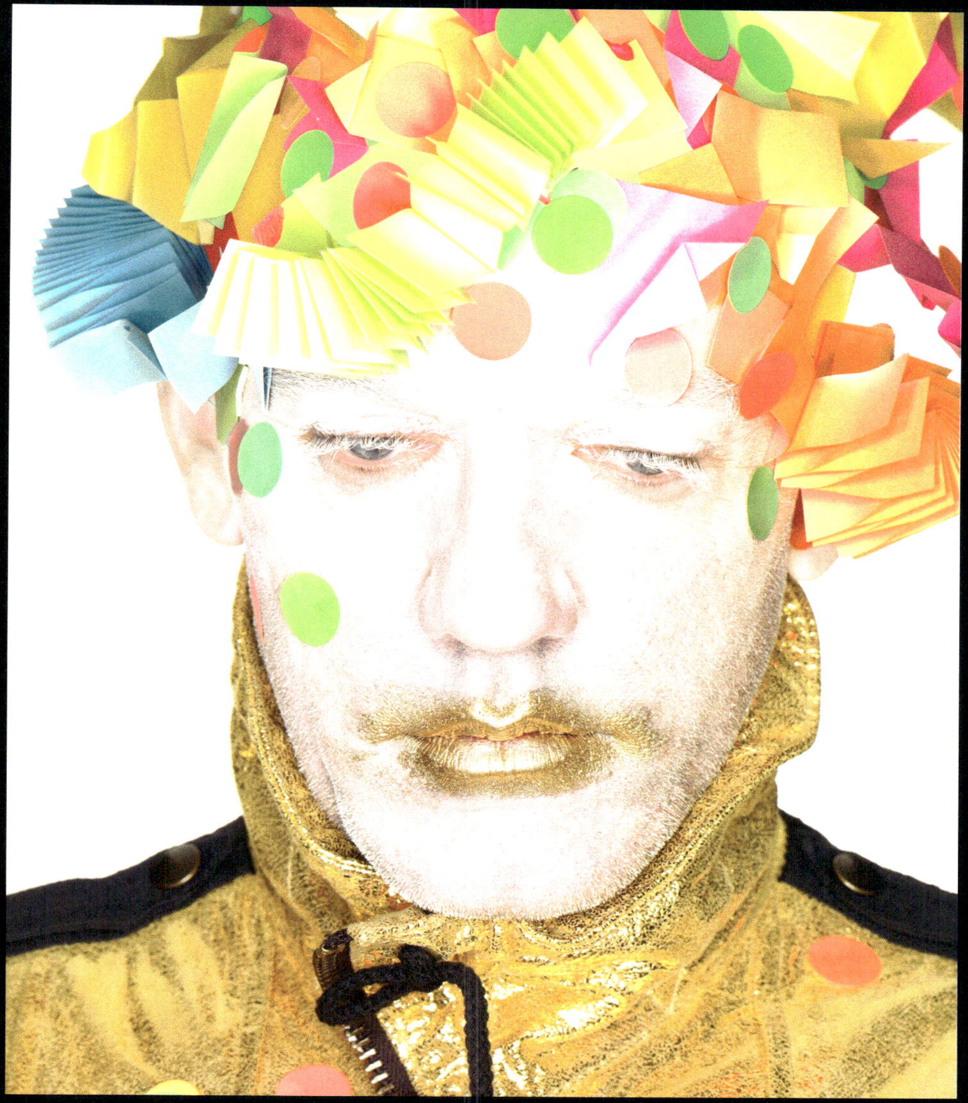

Model
Michael Stipe
Hair Art
Alex Box

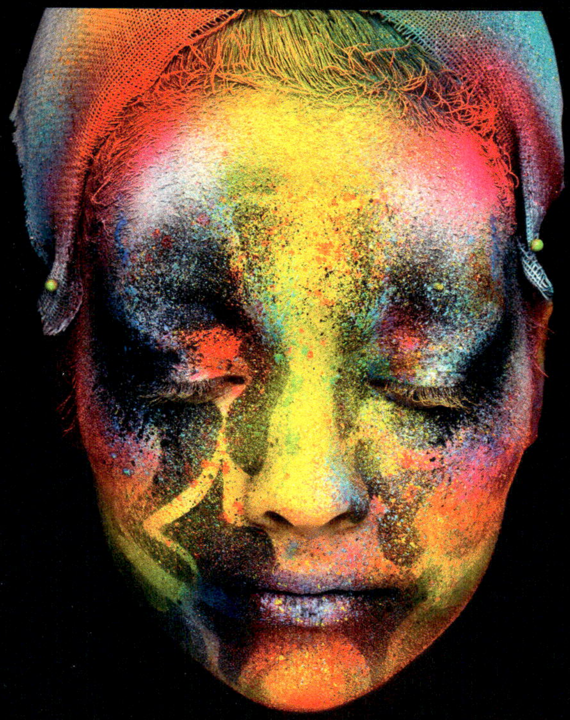

Model
Charly Wright
Hair
James Rowe
at D&V Management

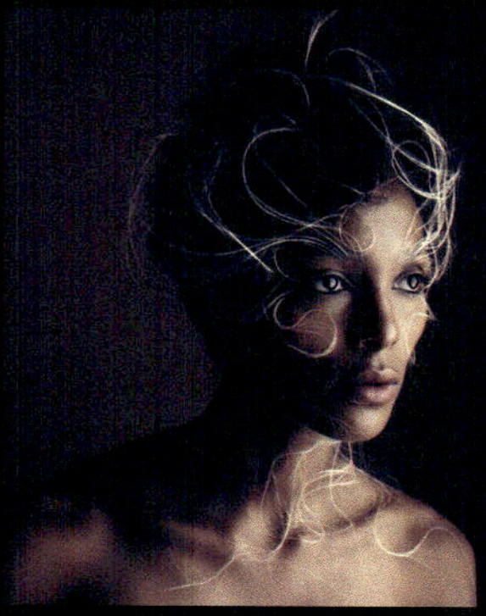

Model
Rhama
Hair
James Pecis
at D&V Management

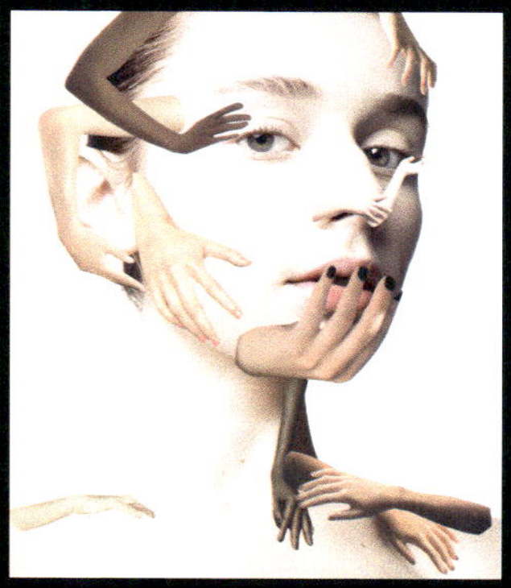

Model
Tatiana
Drawing
Alex Box

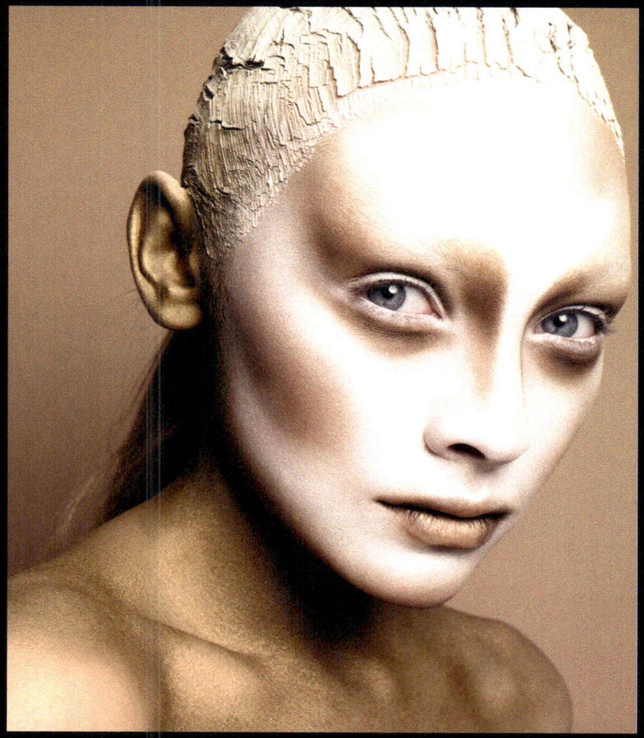

Model
Charly Wright
Hair
Duffy

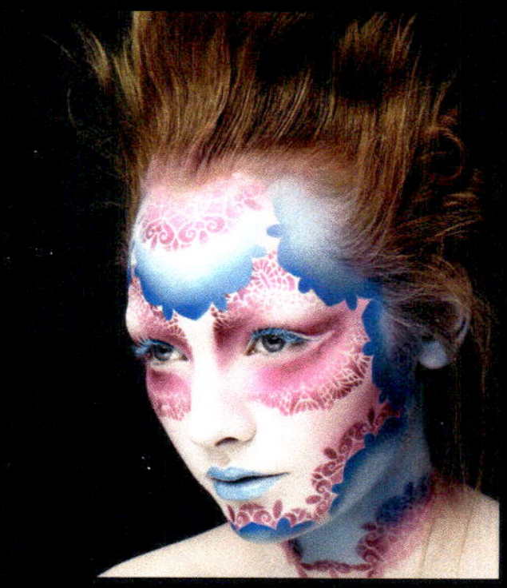

Model
Charly Wright
Hair
James Rowe
at D&V Management

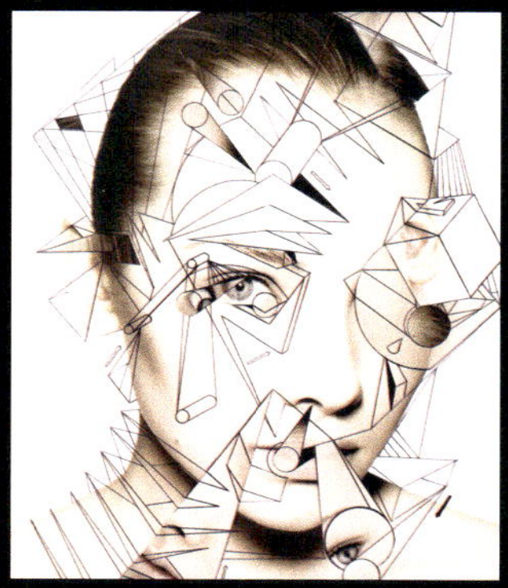

Model
Charly Wright
Drawing
Alex Box

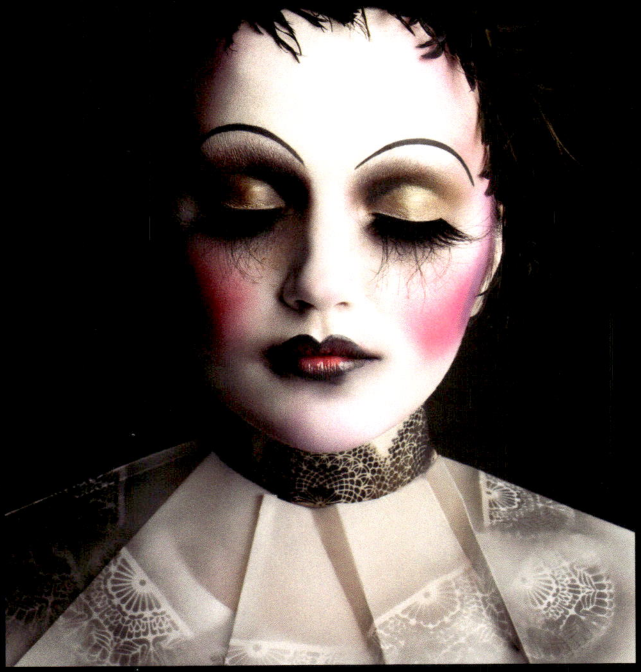

Model
Dangole
Hair
James Rowe
at D&V Management

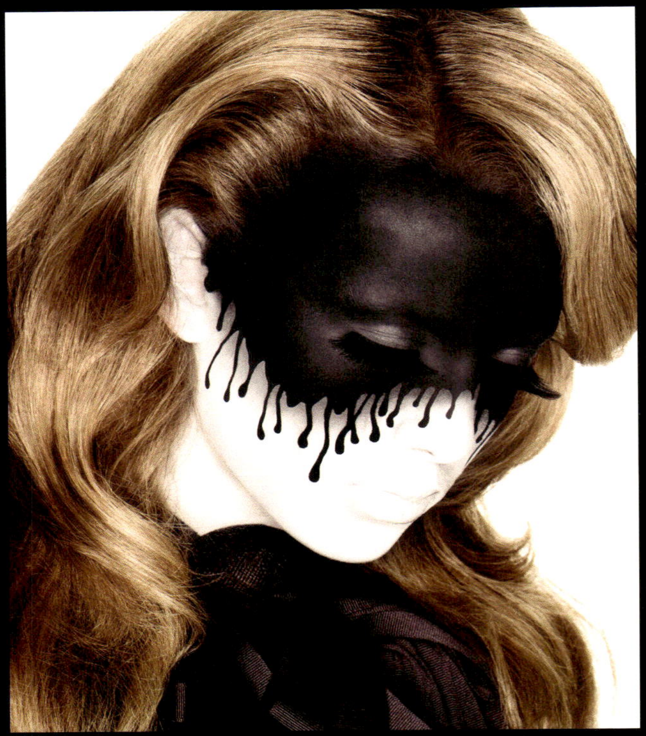

Model
Harriet
Hair
James Rowe
at D&V Management

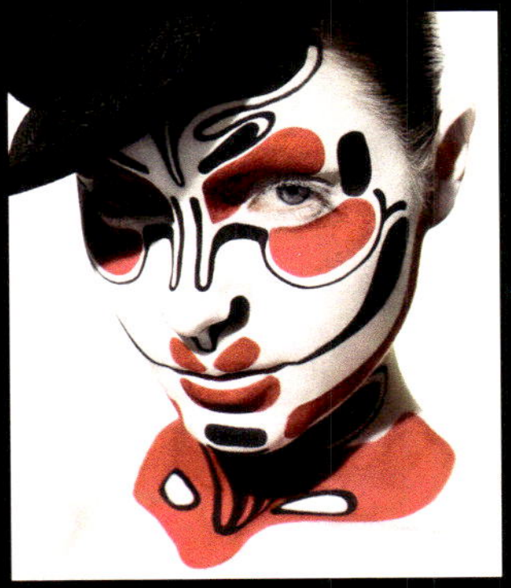

Model
Adrianna Grzadziel
Hair
James Pecis
at D&V Management

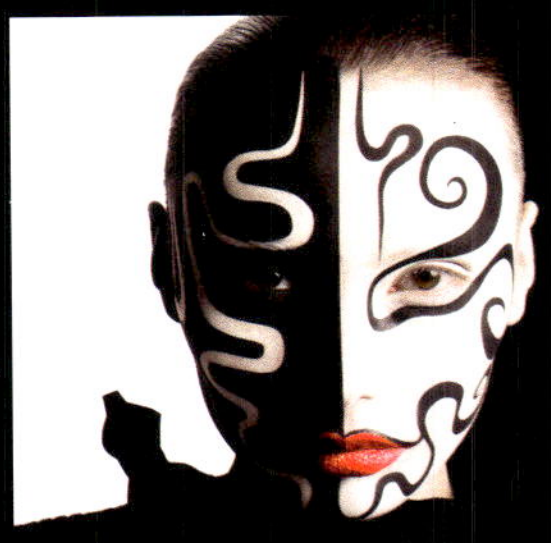

Model
Lakiza

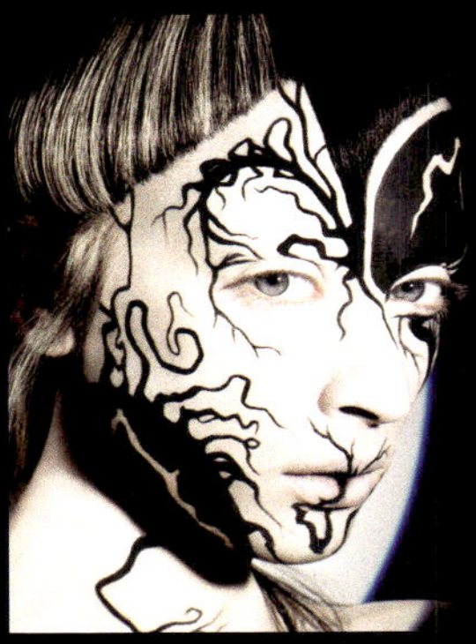

Model
Elina
Hair
James Rowe
at D&V Management

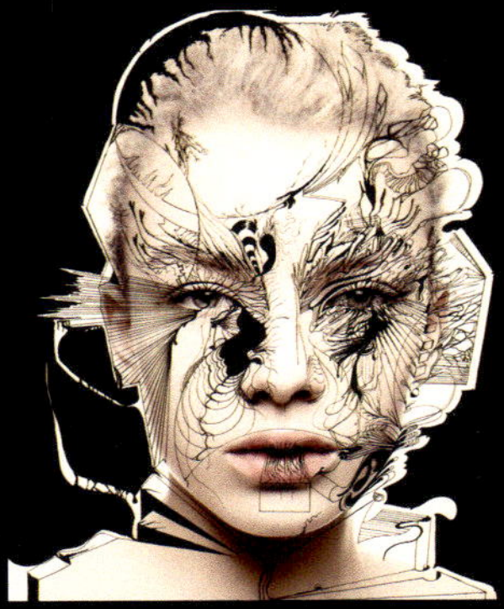

Model
Hattie

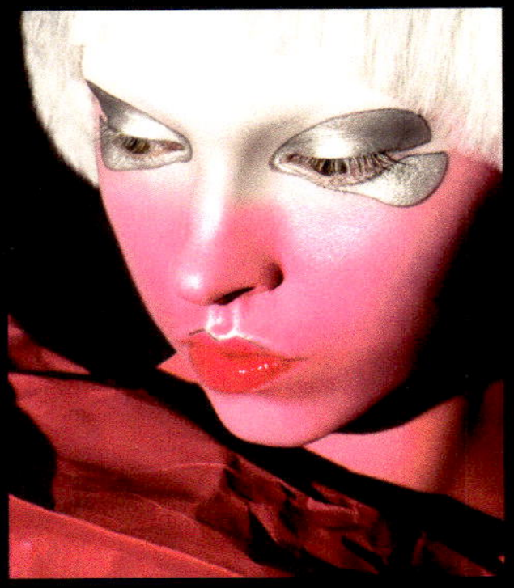

Model
Kinga
Hair
James Pecis
at D&V Management

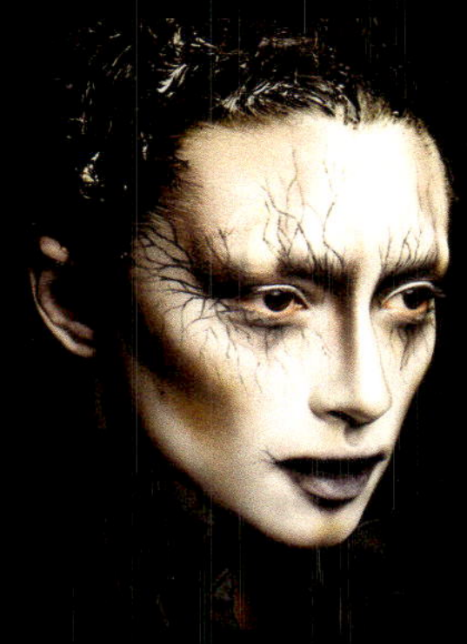

Model
Elina
Hair
Duffy
at D&V Management

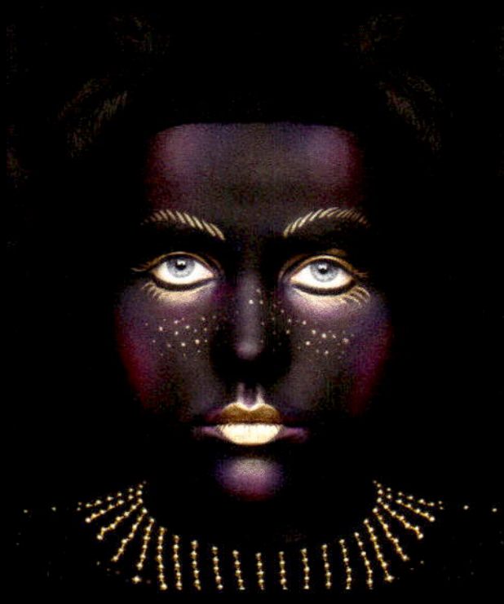

Model
Beta
Hair
James Pecis
at D&V Management

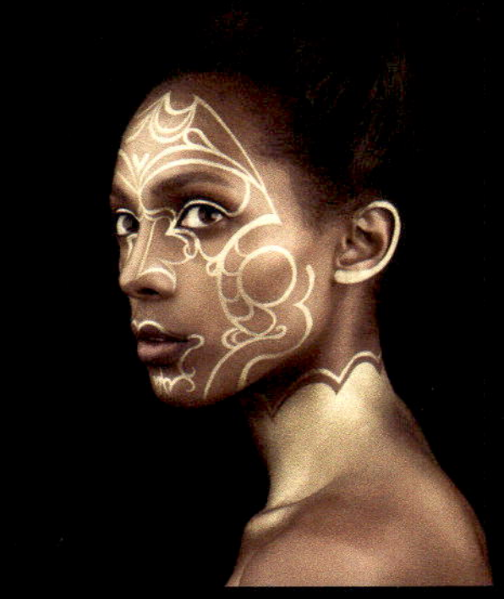

Model
Rhama
Hair
James Pecis
at D&V Management

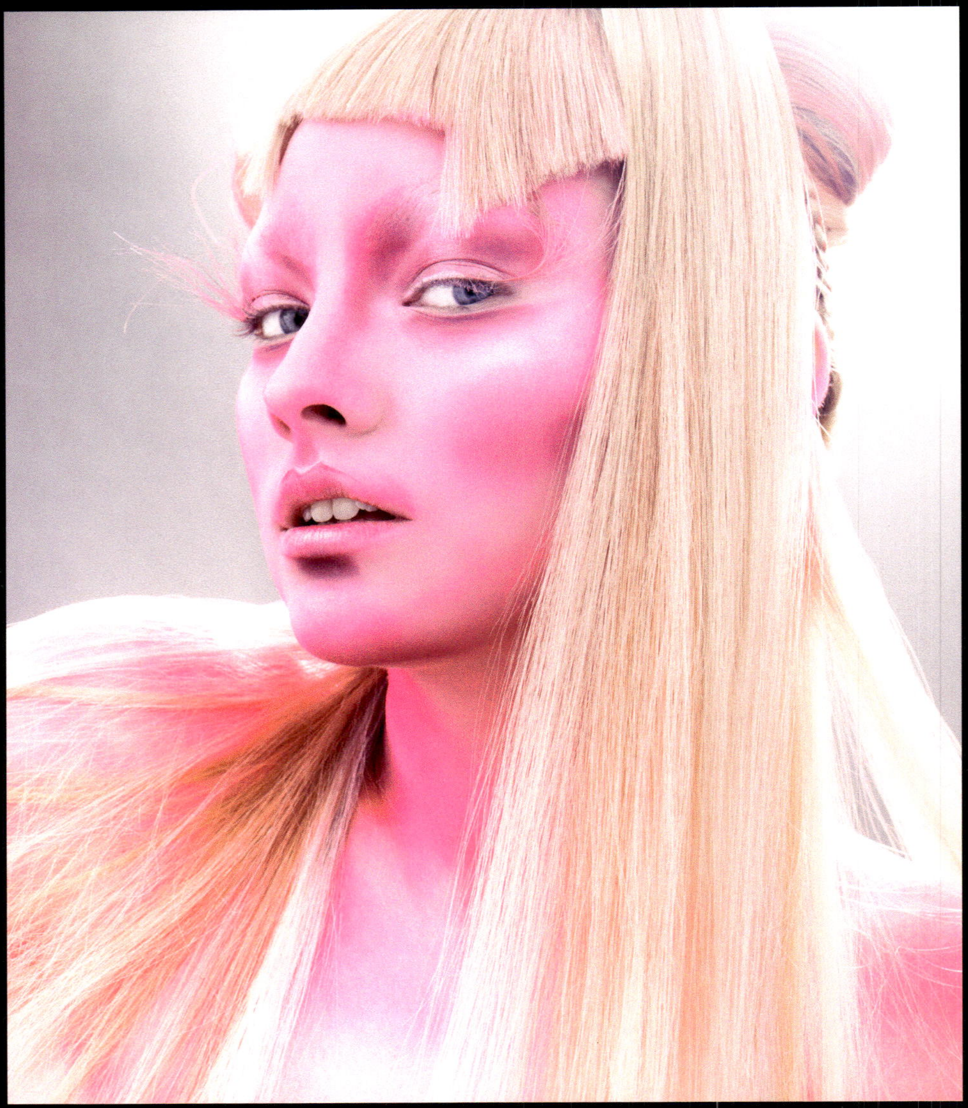

Model
Charly Wright
Hair
James Rowe
at D&V Management

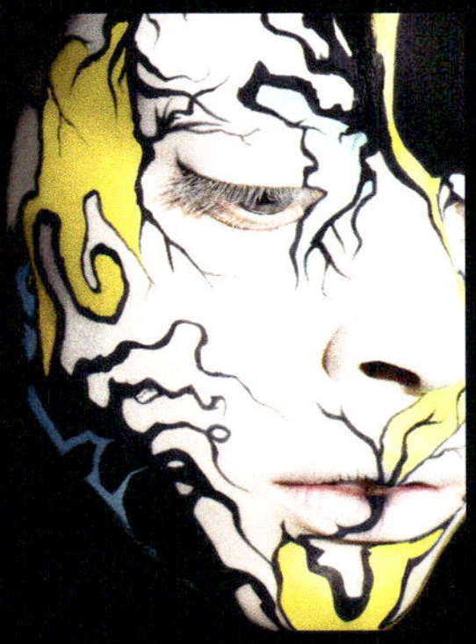

Model
Elina
Hair
James Rowe
at D&V Management

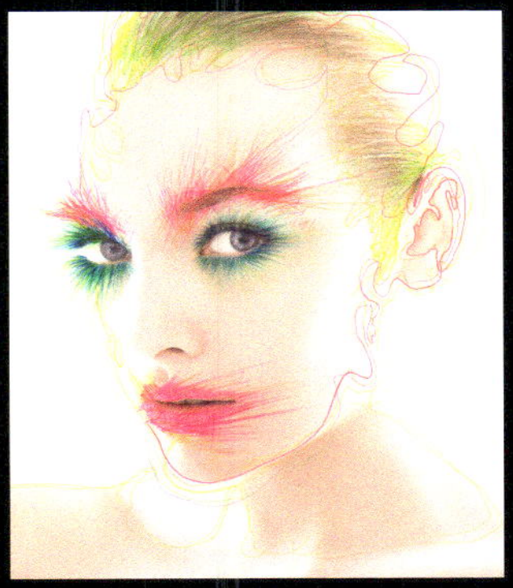

Model
Charly Wright
Drawing
Alex Box

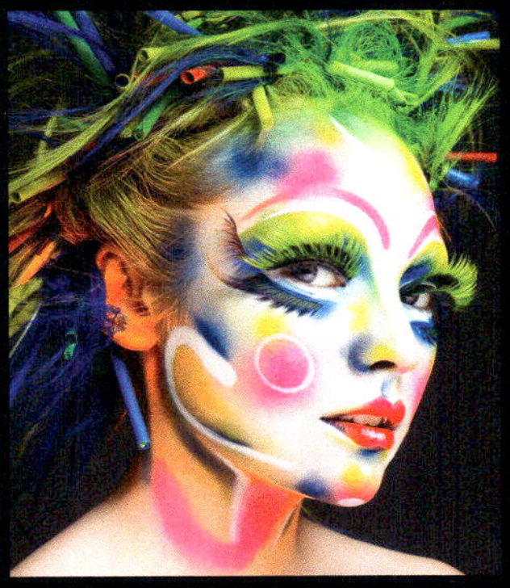

Model
Charly Wright
Hair
James Pecis
at D&V Management

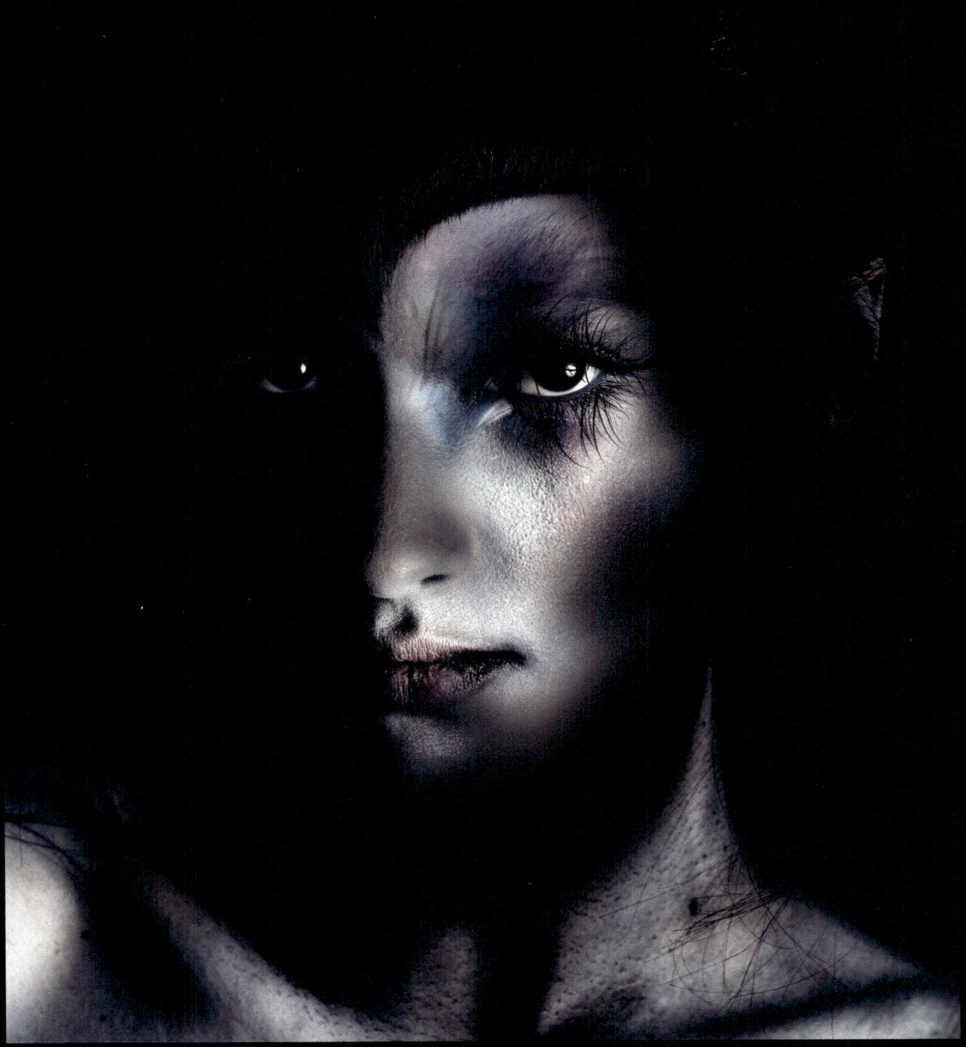

Model
Vesna
Hair
James Rowe
at D&V Management

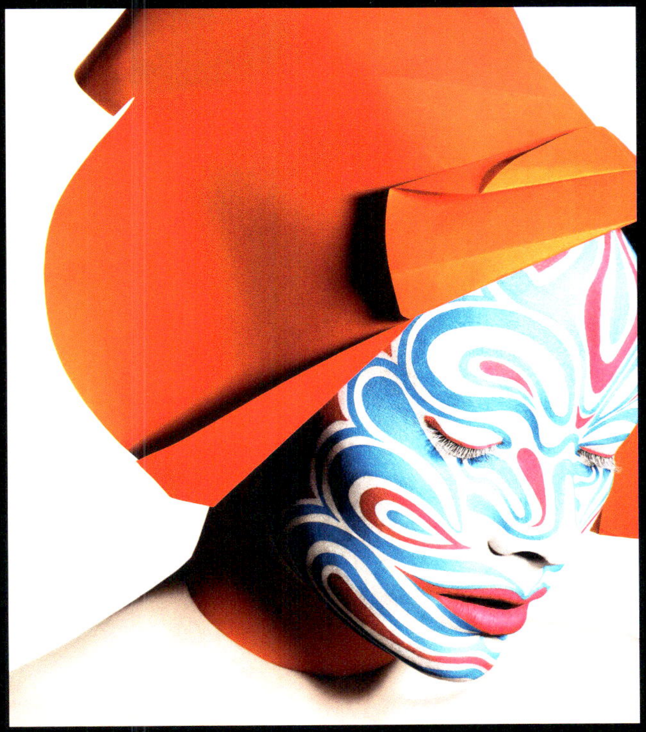

Model
Charly Wright
Hat
Alex Box

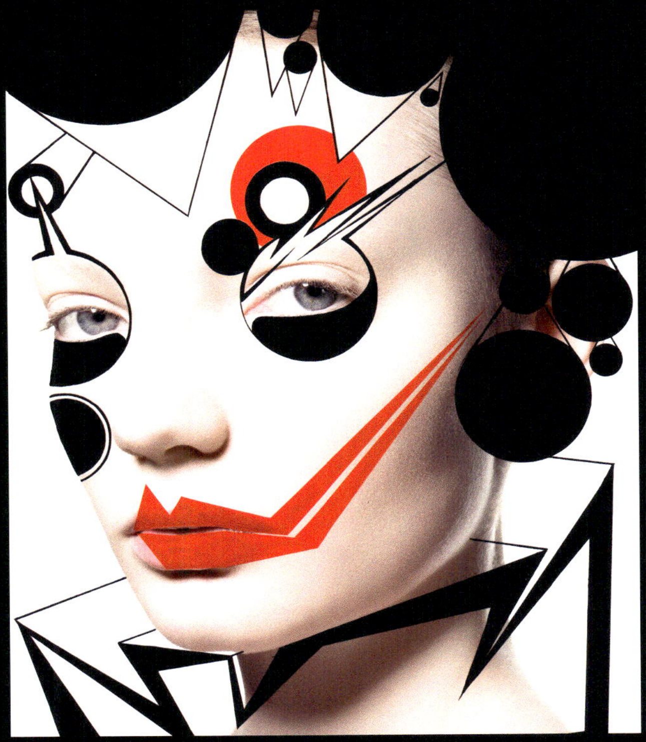

Model
Natasha
Drawing
Alex Box

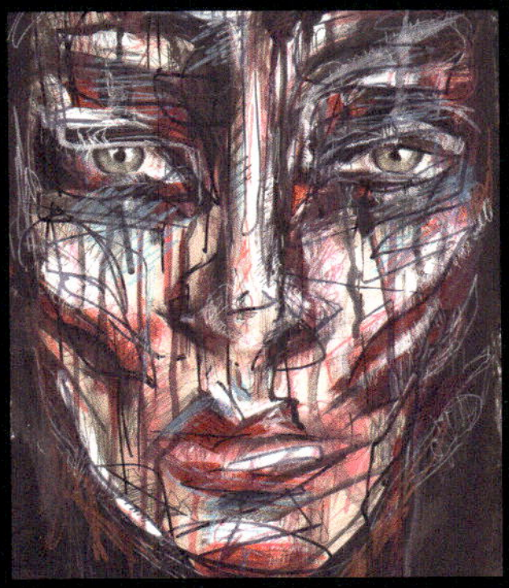

Model
Hattie
Drawing
Matt

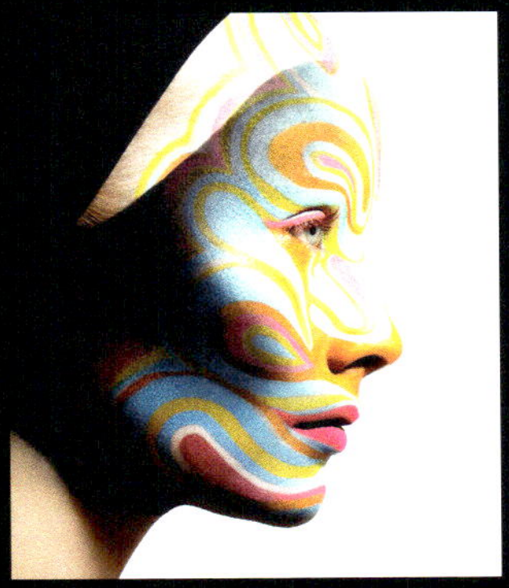

Model
Charly Wright

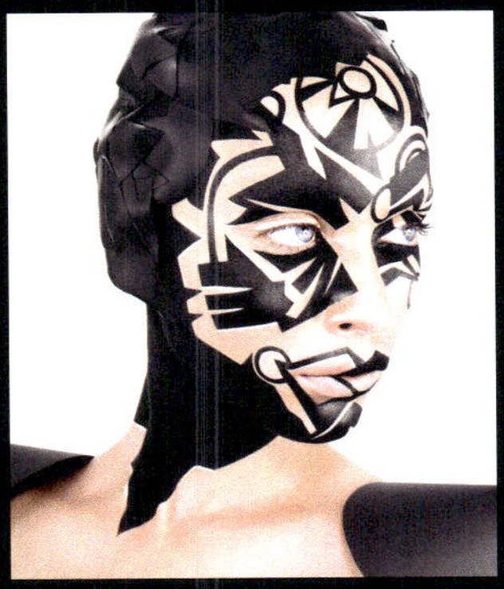

Model
Vesna

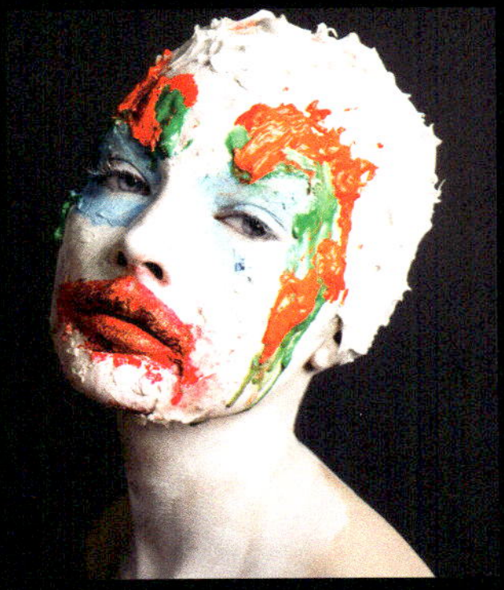

Model
Charly Wright
Hair
James Rowe
at D&V Management

Credits & Thank Yous

Thank-you to
Dee, Celine, Holly, Nancy, James R, James P, Duffy, Steven, all the models, Emma, Thea, Dom, Andrew, Rachel, Triona, Mike, Damien, Paul, Jason, Steve, Bryan & Ryan.

Special thanks to:
Rankin, Charly, Liza, Paul D & Neil.

I strongly second all of Alex's thank you's, especially the ever patient & beautiful Charly. My greatest thanks go to Alex – for her endless flow of ideas and astonishing talent. A unique collaboration that I am sincerely proud to be involved in – Rankin

Design
SEA

Print
Westerham Press

Paper
170gsm heaven 42 softgloss manufactured by Scheufelen, available from Robert Horne. www.roberthorne.co.uk

ISBN 978-0-9563155-0-2

Using ILLAMASQUA make-up

www.illamasqua.com

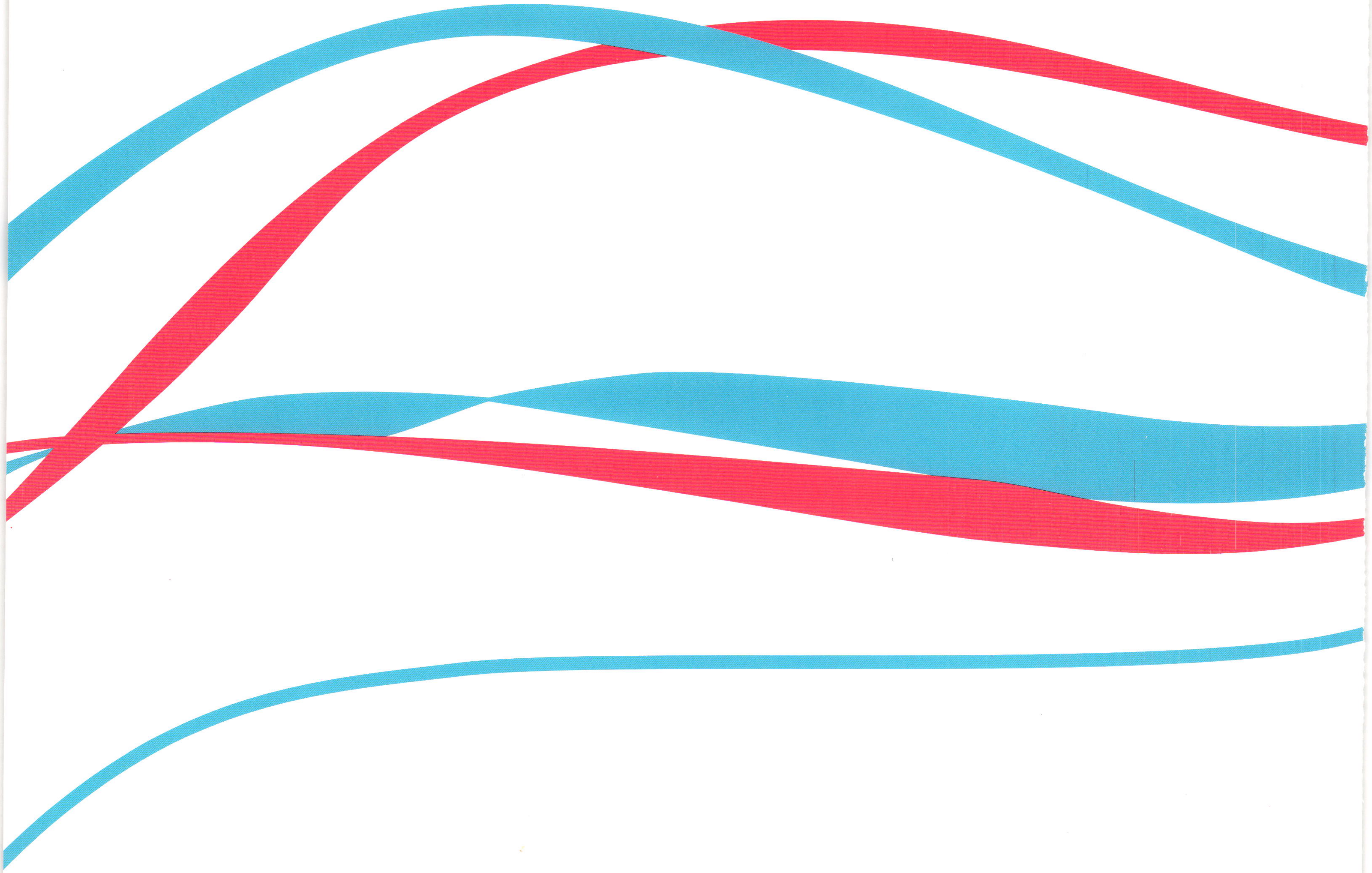